RADIOASSAY IN CLINICAL MEDICINE

RADIOASSAY IN CLINICAL MEDICINE

Edited by

WILLIAM T. NEWTON, M.D.

Professor of Surgery
Washington University School of Medicine
Chief, Surgical Service
Saint Louis Veterans Administration Hospital
Saint Louis, Missouri

and

ROBERT M. DONATI, M.D.

Associate Professor of Medicine
Director of the Section of Nuclear Medicine
Saint Louis University School of Medicine
Chief, Nuclear Medicine Service
Saint Louis Veterans Administration Hospital
Saint Louis, Missouri

CHARLES C THOMAS · PUBLISHER
Springfield · Illinois · USA

Published and Distributed Throughout the World by

CHARLES C THOMAS • PUBLISHER

Bannerstone House

301–327 East Lawrence Avenue, Springfield, Illinois, U.S.A.

ISBN 0–398–03012–X

Library of Congress Catalog Card Number: 73–14898

With THOMAS BOOKS careful attention is given to all details of manufacturing and design. It is the Publisher's desire to present books that are satisfactory as to their physical qualities and artistic possibilities and appropriate for their particular use. THOMAS BOOKS will be true to those laws of quality that assure a good name and good will.

Printed in the United States of America

K–8

Library of Congress Cataloging in Publication Data

Newton, William T.

Radioassay in clinical medicine.

1. Radioisotopes in medical diagnosis. 2. Radioimmunoassay. 3. Hormones—Analysis. I. Donati, Robert M., joint author. II. Title. [DNLM: 1. Hormones—Analysis. 2. Radioimmunoassay. QY330 W567r 1974]

RC78.7.R4N48 615′.7 73–14898

ISBN 0–398–03012–X

CONTRIBUTORS

GAETANO BAZZANO, M.D., PH.D.
Associate Professor of Biochemistry and Medicine
Director of the Section of Nutrition
St. Louis Unversity School of Medicine
St. Louis, Missouri

GAIL S. BAZZANO, PH.D.
Research Associate
Section of Nutrition and Metabolism
Assistant in Medicine
St. Louis University School of Medicine
St. Louis, Missouri

LINCOLN I. DIUGUID, PH.D.
Research Chemist
St. Louis Veterans Administration Hospital
St. Louis, Missouri

ROBERT M. DONATI, M.D.
Associate Professor of Medicine
Director of the Section of Nuclear Medicine
St. Louis University School of Medicine
Chief, Nuclear Medicine Service
St. Louis Veterans Administration Hospital
St. Louis, Missouri

NEIL I. GALLAGHER, M.D.
Professor of Medicine
St. Louis University School of Medicine
Chief, Medical Service
St. Louis Veterans Administration Hospital
St. Louis, Missouri

GARRETT A. HAGEN, M.D.
Associate Professor of Medicine
St. Louis University School of Medicine

Associate Chief, Unit III Medicine
St. Louis Veterans Administration Hospital
St. Louis, Missouri

HELMUT HAIBACH, M.D.
Assistant Professor of Medicine
St. Louis University School of Medicine
Associate Chief, Nuclear Medicine Service
St. Louis Veterans Administration Hospital
St. Louis, Missouri

ROBERT E. HENRY, M.D.
Assistant Professor of Medicine
St. Louis University School of Medicine
Associate Chief, Nuclear Medicine Service
St. Louis Veterans Administration Hospital
St. Louis, Missouri

BERNARD M. JAFFE, M.D.
Associate Professor of Surgery
Washington University School of Medicine
St. Louis, Missouri

WILLIAM T. NEWTON, M.D.
Professor of Surgery
Washington University School of Medicine
Chief, Surgical Service
St. Louis Veterans Administration Hospital
St. Louis, Missouri

CHARLES W. PARKER, M.D.
Professor of Medicine
Washington University School of Medicine
St. Louis, Missouri

FRANCIS A. ZACHAREWICZ, M.D.
Assistant Professor of Medicine
St. Louis University School of Medicine
Educational Coordinator
St. Louis Veterans Administration Hospital
St. Louis, Missouri

PREFACE

THE NUMBER AND VARIETY of chemical compounds of clinical or investigative interest seems almost limitless. Further, many compounds of extraordinary physiologic potency may function in biologic material in extremely low concentrations. Both of these factors combine to confound the ordinary chemical laboratory seeking specific quantitative analyses of these compounds. Fortunately, the advent of radionuclides that can be measured in extremely low chemical concentrations and techniques to attach them to many compounds as trace labels have opened new approaches to quantitative study. However, in spite of a well-informed public regarding the properties of radionuclides and perhaps their use in immunoassay of some protein hormones, the editors of this volume have the impression that there is lack of appreciation of both the variety of approaches and the variety of classes of compounds that can be examined. Consequently, we gathered in this short volume a set of papers chosen to illustrate this variety, yet with each subject presented in sufficient detail to allow the reader to gain insight into the problems inherent to the individual systems.

CONTENTS

RADIOASSAY IN CLINICAL MEDICINE

CHAPTER 1

PRINCIPLES UNDERLYING CURRENT RADIOIMMUNOASSAY TECHNIQUES

WILLIAM T. NEWTON AND BERNARD M. JAFFE

INTRODUCTION

MANY COMPLEX CHEMICAL compounds of biologic importance exist in such low concentrations in biologic fluids and tissues that they do not readily lend themselves to conventional chemical identification and analysis. Development of the techniques of radioimmunoassay in recent years has extended by several orders of magnitude the sensitivity of quantitative measurement of a rapidly growing list of these compounds. For many of these radioimmunoassays have replaced cumbersome and inexact bioassay techniques. The techniques of immunoassay require the production of serum antibodies to bind specifically the chemical complex being estimated. These antibodies can then be used by a variety of maneuvers to identify and quantitate the presence of the chemical complex in a reaction medium. Addition of a radiolabel to the chemical complex allows extension of the measurement of antibody-bound complexes to the very low concentrations that these compounds are found in biologic media.

Each individual radioimmunoassay technique presents individual problems as presented in subsequent chapters of this volume. However, contained principles underlie all of the techniques, and a common thread of empiric findings applicable to a variety of systems is emerging.

REAGENTS

Production of Specific Antibodies

Almost all Vertebrata of the animal kingdom react to the injection into their tissues of foreign proteins or polysaccharides by the generation of an immune response. One of the more frequent manifestations of this response is, of course, the appearance of specific antibodies in the globulin fractions of the serum. And, although the initiation of the immune response requires an immunogen of high molecular weight, the resultant antibodies can be shown to bind specifically only to small areas of the immunogen, no more than five or six amino acids in a chain or adjacent to one another stereoisometrically. Fortunately, for the purposes of immunoassay, chemical covalent conjugation of many chemical compounds to the side chains of an immunogen frequently induces the animal recipient to produce antibodies with specificities to the compound even though it is not itself a peptide or polysaccharide. For example, injection of human serum albumin into rabbits evokes the production of a population of antibody molecules with the capacity to bind to at least three separate amino acid chains of the HSA molecule. Chemical conjugation of 2, 4, dinitro-benzene sulfonic acid to the epsilon amino groups of lysine members of the protein chain of the serum albumin produces an immunogen which stimulates the production of antibodies specific not only for the usual three amino acid chains but also antibodies that bind dinitrophenyl lysine specifically. Production of defined immunogens and analysis of the reactions of the resultant antibodies with defined "antigenic determinants" have formed the basis of a large portion of modern immunochemistry.

Preparation of Immunogens

A variety of natural proteins and even chemically produced random copolymers of one to three amino acids have been used as "backbone" carrier molecules. In order to elicit specific antibodies animals are immunized with an immunogen in which the chemical complex to which antibodies are desired is conjugated to side reactive groups of the carrier molecule. Of course, some important biologic molecules are large enough without conjugation to induce immune responses (growth hormone, insulin) or can be isolated from biologic sources already bound to immunogenic molecules (crude hog gastrin). If conjugation is necessary almost all immunogenic proteins contain reactive side groups available for conjugation. The gamma carboxyl group of glutamic acid and the epsilon amino group of lysine provide sites for peptide and azo linkages. The reader is referred to standard texts on amino acid and peptide chemistry for the variety of reactions available as well as to the individual techniques described in this volume. Choice of techniques will be dictated in large measure by the type of reactive group on the native biologic compound or the type of reactive group that can be conveniently introduced on this compound. Direct ethereal or peptide linkages are to be preferred, but use of intermediary linkage compounds such as imidoesters,[4] toluene diisocyanate[5] or bis diazobenzidine may be necessary. The most popular reaction recently has become the direct peptide linkage of either amino groups or carboxyl groups of the target compound or some derivative to the protein through the intermediary reaction with carbodiimide compounds.[6] The reader should be aware that all of these reactions introduce intermediary compounds and that antibodies specific for the intermediary will be found in resultant antiserum. Even the carbodiimides form immunogenic radicals, and binding of antibodies to these conjugates may be specific for the carbodiimide used as well as the compound conjugated. Fortunately, the two water soluble carbodiimides that are readily available do not produce cross-reactive antibodies.[7] Therefore, one carbodiimide can be used to synthesize the immunogen leaving the other available for synthesis of target compounds to be used in the reaction mixture if such a synthesis is needed.

In general, the more intrinsically immunogenic the carrier molecule is the more likely are conjugates of target compound to that molecule to induce antibodies to the conjugated compound. And, there appears to be little or no advantage to using a carrier molecule that excites few antibodies to its own structure. Although immunization with a conjugate of, say bradykinin molecules to a relatively poor immunogenic carrier such as polylysine may produce a more "pure" antibody population to the bradykinin groups, the total response is apt to be weak and demonstrable in only a few of the animals immunized.[8] As pointed out below the specificity of the reaction can usually be controlled precisely by manipulation of the reaction system and control of the target radiolabeled compound. The reader should realize that this point is controversial, and there can be envisaged circumstances where a host of interfering and cross-reacting substances in normal serum require the production of a more pure antibody population. Random copolymers of lysine, alanine and glutamic acid are available and are moderately immunogenic in themselves. They contain convenient side chains for conjugation reactions.

Another approach to the purity problem may be available by completely saturating the side chains of the immunogenic carrier molecule with the compound to be assayed. If bovine gammaglobulin, a highly immunogenic protein, is reacted with large excesses of dinitro-benzene sulfonic acid to the extent that sixty dinitrophenyl groups are conjugated per molecule BγG to the available lysine amino groups, the resultant conjugate is still highly immunogenic, but the antibodies produced are almost all specific for dinitrophenyl groups, and antibodies to the native BγG antigenic determinants may be absent or difficult to demonstrate.

Immunization Procedures

In general, the longer the time interval from initial injection to harvest of antiserum the broader will be the specificity of the antibody population. And, in the usual circumstance where the compound to be assayed forms only one of many antigenic determinants presented to the immunized animal it might be predicted that prolonged immunization times would be required. And, indeed this is the case with antibodies specific for the compound which are usually found only after three or four months and frequently longer. Further, intensive immunization schedules with several secondary immunization injections have usually been necessary. Most investigators have also used adjuvant techniques such as Freund's incorporation of immunogen into a water-in-oil emulsion for injection or alum precipitation of immunogen.

Although a desirable immunization dosage might be in the range of one milligram immunogen per kilogram body weight per injection, these quantities are frequently out of the question due to the expense or rarity of the compound. In these circumstances it seems wisest to space the limited quantity available over several monthly injections rather than to commit all immunogen to a single injection.

The choice of animals to be immunized will depend to some extent on the volume of antiserum needed for the proposed investigations. A single guinea pig may provide enough antiserum for several years' work in some situation, since assays can frequently be done with 1:10,000 dilutions of antiserum (see below).

It would seem reasonable that the more disparate phylogenetically the animal is from the biologic source of the material to be assayed, the more foreign would be the immunogen and more likely that an immune response would ensue. When attempts to immunize rodents to human parathormone failed, chickens were found to be suitable responders to immunization.[9] Since the ability to respond to immunization is genetically determined, it would seem wise to use several mongrel animals rather than members of a given genetic strain. Of course, the animals chosen should be known to be capable of prolonged healthy survival in the facilities available.

REACTIONS

The Reaction Medium

Before attacking the problem of analyzing the reaction of antigen and antibodies *in vitro,* brief consideration should be given to the milieu in which the reaction takes place. Since the bonds of the complexes are relatively weak, they are sensitive to the alterations that affect hydrogen bonds. Alteration of pH from neutrality must be guarded against, and ionic strength should be low, in the range of physiologic saline. Although reactions take place rapidly at 37° C, complexes are more stable in the cold.

Care must be taken that the medium does not allow chemical or physical alteration of the reactants. Many biologic fluids contain potent lytic enzymes for a variety of chemical bonds. Addition of epsilon amino caproic acid, an inhibitor of many proteolytic enzymes found in serum, to systems analyzing reagents with peptide bonds is frequently helpful in stabilizing these systems. Anti-oxidants or other reagents may be required. It may be necessary to separate the chemical to be estimated from its biologic source by one of the various chemical extraction procedures before a stable analysis system can be set up.

Many systems seem to require other large molecules in the medium to give most reproducible results. In an anti-peptide system the authors found egg albumin in concentrations of 2.5 mg per ml superior to ficoll, hydroxyethyl starch, dextrans and other proteins. Individualization of each assay system is part of the art of immunoassay and cannot be neglected.

Reaction of Antibodies with Antigens

The reactions, *in vitro,* of antibodies with antigens and particularly with defined chemical determinants, termed haptens, have been intensively studied. In a representative system antibodies raised to an immunogenic protein to which dinitrophenyl groups have been conjugated are separated and purified from the harvested antiserum by one of a variety of techniques. These antibodies are then examined in their reaction with the specific hapten, dinitrophenyl lysine, or other dinitrophenyl compounds. Techniques exist for the separation of bound and unbound hapten from the reaction systems and allow the construction of Sip's plots or other mathematic manipulations. The reader is referred to standard immunochemistry texts for more detailed treatment of these reactions.

Immunochemical experimentation has shown that the usual IgG antibody molecule of about 150,000 average molecular weight contains two binding sites. Thermodynamic treatment of the data shows that the binding forces of sites for specific haptens have free energy exchanges in the order of those found for hydrogen bonds. Despite the low binding energies the intrinsic association constants reflect a high degree of completion of the reaction of antibodies with hapten to form complexes. These binding constants are developed from the law of mass action much as for any other chemical reaction and, in practical terms for this discussion, represent the molar concentration of hapten at which half saturation of the antibody binding sites occurs (the number of bound and unbound hapten molecules are equal). Values of 1×10^{-5} M are easily obtained, and population of antibodies with values of 5×10^{-9} M are not rare. Thus, serum antibodies are excellent reagents for the binding of many chemical groups in theory. For the purposes of assay it is now required that means be available to determine that this reaction has occurred and to what degree.

Measurement of Complex Formation—Indicator Techniques

Historically, the oldest immunoassay techniques employed a visual indicator of the reaction of antigen and antibodies. In these techniques the antigenic determinant or whole antigen was absorbed to or covalently joined to a visible particle such as the human erythrocyte. Since antibodies are multivalent, their union to antigenic determinants on separate red cells could lead to the complex process of hemagglutination with separation from the uniform erythrocyte suspension of large clumps of cells. Latex particles have also been used for this purpose. Quantitatively, the reaction is measured by titer, the highest dilution of antibodies that produces visible agglutination. Once the basic system is established, varying known quantities of unconjugated antigenic determinants could be added to the system to inhibit the agglutination by competing with the particle bound antigen for antibody binding sites. A standard inhibition curve with the titer of antibodies as a function of the amount of added inhibitor can be set up. Measurement of the antibody titer after the addition of an unknown sample of biologic material compared to the titers obtained after the addition of known amounts of the specific chemical allows determination of the amount of the specific chemical in the sample. The indicator techniques are cumbersome (production of stable antigen-particle conjugates) and have been replaced almost entirely by the separation techniques discussed below.

Measurement of Complex Formation—Separation Techniques

Measurement of the amounts of antibody bound compound and the free unbound compound at equilibrium in a given reaction can be simplified markedly if the two reaction products can be separated physically and analyzed individually. A variety of techniques have been developed for this purpose and are discussed below. It is, of course, important that the separation process not alter the equilibrium established. Consequently, most separations are carried out in the cold. Very little attention has been paid to this point in the literature. For example, gel filtration columns have been extensively used to separate out the small chemical groups and allow the larger bound antibody-chemical complexes to percolate down the column. As the mixture of reactants proceeds down the column the concentration of unbound simple chemical might fall altering the equilibrium of the reactants and, theoretically at least, causing additional simple chemical groups to separate from the antibodies. In some studies with peptide hormones as the target chemical this dis-equilibrium has not seemed to be a significant problem, but this might not be the case for smaller molecules or different gels. Of course, the separation process should not produce chemical or physical alterations of either of the two reactants that affect their ultimate analysis. Mention of this requirement has been made above in describing the milieu of the reaction mixture; care should be taken that these precautions are carried through the separation process.

Of the various separation processes, the direct precipitation of antibody-bound complexes from the reaction medium is the simplest in theory. Most antibodies readily form precipitable complexes when their specific antigens are part of proteins or similar large molecules. A now classic example would be the precipitation of ovalbumin by rabbit antibodies. However, the conditions required for accurate and reproducible analyses are rigorous, slow and inconvenient for large scale work. Although specific precipitation is not used widely today for immunoassays, the basic principles form the foundation for immuno diffusion analyses which have been widely used for estimation of concentrations of serum proteins. None of the direct precipitation methods allows analyses much below the microgram per milliliter level and are, therefore, not suited to the very low concentration of many important compounds found in biologic fluids.

On the other hand, a variety of coprecipitation processes have been used successfully for radioimmunoassays. The antibody target chemical complexes can be precipitated at neutral pH without disrupting the complex in high concentrations of sodium or ammonium sulfate salts. Although these processes may increase the sensitivity of measurements a thousand-fold over that obtained by direct precipitation, there is a slight solubility of the complexes that prevents most systems from reaching the picomole per milliliter level needed for many compounds. Of more value have been the so-called "double antibody" coprecipitations in which the complexes of antibodies and simple chemicals is precipitated by addition of antibodies to the serum globulins that form the antibodies of the complexes.[10] In a representative system antibodies are raised to human bradykinin in rabbits, and these antibodies are used to react with bradykinin in the test system. Addition of antibodies raised in goats by immunization with rabbit IgG globulin causes precipitation of the rabbit globulin antibodies and coprecipitation of the bradykinin. Frequently, it is necessary to add additional rabbit IgG globulin to the system before the goat antibodies, and to increase the amount of goat antibodies correspondingly (calculated from separate precipitation analysis of the reaction of goat antibodies with rabbit IgG). This additional precipitate need be only of the order of a few micrograms of rabbit IgG in order to form enough precipitate to insure quantitative precipitation of the immune complexes. The additional precipitate increases the nonspecific inclusion of unbound chemical in the total mass, and the proper addition that can be tolerated must be determined experimentally. Nonspecific binding of complexes of antibodies and target chemicals to charcoal particles coated with dextran has been used to separate bound and unbound chemical. In some cases simple binding of antibodies to the plastic or glass walls of the reaction tube is of sufficient binding energy as to allow the segregation of target molecules from a fluid system. Other solid phase techniques to bind antibodies chemically or physically can be thought of, and some of these have been successful.

Another group of separation procedures are those based on the presence of electrical charges on antibody molecules. Anion or cation exchange resins can be manipulated in either column or batch techniques to bind antibody molecules securely together with their complexed chemical group. Migration of antibody complexes in an electrical field is limited at neutral or slightly alkaline pH, and many unbound chemicals readily separate from the mixture and migrate at more rapid rates. These electrophoretic techniques involve a great deal of nonspecific binding by the supporting medium and are more useful for identification procedures than for quantitative estimation of concentrations.

Use of gel filtration to separate complexes and unbound chemicals has also been applied frequently to immunoassay problems. The gel materials are composed of polymeric chains as in polyacrylamide or dextrans which are subjected to cross-linking reactions to produce chemical lattices with the open spaces of the lattice determined by the degree of cross-linking. Passage of a group of molecules down a column of hydrated gel results in entrapment of the smaller molecules which can penetrate the lattice and the unhindered passage of molecules too large to be engaged in the lattice. By a proper choice of gel material a system can usually be set up that allows separation of the smaller unbound chemical from the antibody complexes of a reaction mixture. Mention was made at the beginning of this section of the potential problem of reaction disequilibrium that might be encountered in this technique. Use of the gels also requires a certain amount of experience in the mechanics and arts of preparing the materials, particularly those with loose lattice linking. Column operations require fractional collection and separate analyses of the eluate fractions. Batch operation is limited by the nonspecific binding of both reactants in the system.

Use of a particular separation process will be determined by the requirements for sensitivity of the assay, but the wide variety of procedures available will allow the investigation to pick the process best suited to his experience.

Radiolabels

Sensitivity of measurement of the reaction of antibodies and a target chemical of the reaction complex that can be separated for analysis can usually be increased by several orders of magnitude if the target chemical can be radiolabeled. Detection and quantitation of extremely small quantities of material is frequently possible. Consider a physiologically potent small peptide of molecular weight 2000 which is present in serum in nanomolar (one picomole per ml) amounts. It would be convenient if 0.1 ml serum containing 0.1 picomole could have an activity of 10,000 cpm representing only 25 mCi per micromol. This specific activity is easily obtainable if there is a suitable group in the peptide for direct iodination. Many compounds of this or even higher specific activity may be purchased from nuclear chemical suppliers, and manipulations may be performed to multiply specific activity even further (*vide infra*).

Choice of an isotope label is usually dictated rather than chosen. Gamma emission has significant advantages over beta or other emission: high counting efficiency over low, long half-life over short, a label intrinsic to the molecule over an added extrinsic label, inexpensive over expensive. Gamma-emitting isotopes of the atoms commonly found in organic molecules are not available in forms suitable for assay work. The gamma-emitting carbon oxygen isotopes that are available from the cyclotron have half-lives far too short for practical use in assay work. The oxygen 15 isotope produced by dueteron in-neutron out bombardment of nitrogen emits positrons which, in turn, produce two high energy gamma rays, but the half-life is only 2.2 minutes. However, the conveniences of handling gamma emission are so great that other gamma-emitting atoms have been used extensively as labels in immunoassay work. By far the largest applications have used the iodine 131 and iodine 125 isotopes, since these can be obtained cheaply in carrier-free state and react readily with protein or peptide tyrosinyl or histidyl groups to form a stable label. The iodine 125 isotope is the more frequently used today because of its longer half-life and perhaps because its lower energy produces less radioautolysis of the target chemical. Should the material to be assayed lack groups for iodination, Goodfriend et al.[11] have described a procedure for the introduction of tyrosinyl groups to the N-terminal end of angiotensin with subsequent iodination to high specific activity. The authors [12] have described the conjugation of a large copolymer of alanine, glutamic acid and tyrosine to a small tetrapeptide to be assayed. This maneuver had the multiplier effect of providing four tyrosine groups for each peptide unit, and the subsequent iodination of the complex produced extremely high activity relative to the peptide.

Iodination requires a brief oxidation step which may destroy the target chemical or so alter its structure as to become unreactive to antibodies. For example, we have been unable to iodinate the prostaglandins for assay (see Chapter 3). And, in rare cases, the simple introduction of a large iodine atom into the target chemical may interfere with antibody binding. For these reasons and others the investigator may choose a beta-emitting isotope for the label. With the exception of phosphorus 32, most atoms that comprise biological compounds have isotopes of low energy emission. Counting these isotopes in decent counting efficiency almost requires liquid scintillation systems. Here, the investigator is confronted with the relative expense of liquid scintillation counting equipment and sample preparation as well as the problem of dispersing hydrophilic biologic compounds in the nonpolar scintillation medium. Fortunately a number of all-purpose proprietary solvents have been developed for this purpose. Alternatively, the investigator may draw on his own experience with solvents such as dimethyl formamide, dimethyl sulfoxide, dioxane or tetrahydrofuran in which biologic materials may be dissolved before being added to the toluene-scintillator solution. The reader is referred to texts dealing with details of the variations available.

If all fails the material may be suspended rather than dissolved in a gel of the scintillation fluid with surprising counting efficiency and reproducibility. Newton et al.[13] found that Ba $^{35}SO_4$ precipitates in the range of 100 to 150 mg washed with dimethyl-formamide could be suspended in gel with counting ef-

TABLE 1–I
TECHNIQUES OF GASTRIN IMMUNOASSAY *

Investigators	*Immunogen*	*Animals*	*Route of Immunication*	*Separation*	*Serum pg/ml Sensitivity*	*CCK-P2 Cross Reactivity*
McGuigan [14,15]	pure human gastrin (SHG) conjugated to bovine serum albumin (BSA)	rabbits	SQ in Freund's	Double antibody	5	0.001
Thompson et al.[16-18]	semipurified hog gastrin + pure human gastrin (SHG)	rabbits	SQ in Freund's	Double antibody	2,500	0.00012
Yalow & Berson [3]	semipurified hog gastrin	guinea pigs	SQ in Freund's	Anion exchange resin	1	0.00027
Yip & Jordan [19]	semipurified hog gastrin conjugated to (BSA)	rabbits	SQ in Freund's	Double antibody	10	0.00001
Jeffcoate [20]	semipurified hog gastrin attached to latex particles	chickens	IM, SQ & IV	Ethanol precipitation	100	0.00001
Ganguli [21]	semipurified hog gastrin	rabbits	SQ in Freund's	Ammonium sulfate & double antibody	30	0.0001
Labo [22]	pure human gastrin conjugated to (BSA)	rabbits	SQ in Freund's	Double antibody	5	—
Hansky & Cain [23]	pure human gastrin conjugated to (BSA)	rabbits	IP, IM & SQ	Dextran-coated charcoal	5	0.0001
Young et al.[24]	gastrin pentapeptide conjugated to (BSA)	chickens	SQ in Freund's + IM H. pertussis	Anion exchange resin	50–100	—

* Reprinted with permission of the publishers from B. M. Jaffe, In *The Practice of Surgery: Current Review*, W. F. Ballinger, and T. Drapanas, eds. (St. Louis, Mosby, 1972).

ficiencies of 70 percent with excellent reproducibility. Developing new liquid scintillation systems is somewhat of an art acquired through experience.

ASSAY SYSTEMS AND RESULTS

From the foregoing pages it should be apparent that a variety of approaches to the development of an assay system are available. The ultimate tests of a system are its specificity and its sensitivity. Table 1–I lists some of the approaches to the assay of the peptide hormone, gastrin and the results of the approach in terms of sensitivity and specificity. The following chapters of this volume have been chosen to illustrate the choice of variables that may be selected by an investigator as best designed to meet his own needs.

REFERENCES

1. Webb, T., and Lapresle, C.: Study of the absorption on and desorption from polysterene-human serum albumin conjugates of rabbit anti-human serum albumin having different specificities. *J Exp Med, 114:*43, 1961.
2. Eisen, H.N., Belman, S., and Carsten, M.E.: Interaction of dinitrobenzene derivatives with bovine serum albumin 1, 2. *J Am Chem Soc, 75:*4451, 1953.
3. Yalow, R.S., and Berson, S.A.: Radioimmunoassay of gastrin. *Gastroenterology, 58:*1, 1970.
4. Dutton, A., Adams, M., and Singer, S.J.: Bifunctional imidoesters as cross-linking reagents. *Biochem Biophys Res Commun, 23:*730, 1966.
5. Schick, A.F., and Singer, S.J.: On the formation of covalent linkages between two protein molecules. *J Biol Chem, 236:*2477, 1961.
6. Sheehan, J.C., and Hlavka, J.J.: The use of water soluble and basic carbodiimides in peptide synthesis. *J Org Chem, 21:*439, 1956.
7. Jaffe, B.M., Newton, W.T., and McGuigan, J.E.: The effect of carriers on the production of antibodies to the gastrin tetrapeptide. *Immunochemistry, 7:*715, 1970.
8. Spragg, J., Schroder, E., Steward, J.M., Austen, K.F., and Haber, E.: Structural requirements for binding to antibody of sequence variants of bradykinin. *Biochemistry, 6:*3933, 1967.
9. Reiss, E., and Canterbury, J.: A radioimmunoassay for parathyroid hormone in man. *Proc Soc Exp Biol Med, 128:*501, 1968.

10. Skom, J.H., and Talmage, D.W.: Nonprecipitating insulin antibodies. *J Clin Invest, 37*:783, 1958.
11. Goodfriend, T.L., and Ball, D.L.: Radioimmunoassay of bradykinin: Chemical modification to enable use of radioactive iodine. *J Lab Clin Med, 73*:501, 1969.
12. Newton, W.T., McGuigan, J.E., and Jaffe, B.M.: Radioimmunoassay of peptides lacking tyrosine. *J Lab Clin Med, 75*:886, 1970.
13. Newton, W.T., Murphy, J., and Mullins, L.E.: Determination of radiosulfate in biologic fluids and tissues. *J Lab Clin Med, 69*:518, 1967.
14. McGuigan, J.E.: Immunochemical studies with synthetic human gastrin. *Gastroenterology, 54*:1005, 1968.
15. McGuigan, J.E., and Trudeau, W.L.: Studies with antibodies to gastrin. *Gastroenterology, 58*:139, 1970.
16. Odell, W.D., Charters, A.C., Davidson, W.D., and Thompson, J.C.: Radioimmunoassay for human gastrin on an antigen. *J Clin Endocrinol Metab, 28*:1840, 1968.
17. Charters, A.C., Odell, W.D., Davidson, W.D., and Thompson, J.C.: Gastrin: Immunochemical properties and measurement by radioimmunoassay. *Surgery, 66*:104, 1969.
18. Charters, A.C., Odell, W.D., Davidson, W.D., and Thompson, J.C.: Development of a radioimmunoassay for gastrin. *Arch Surg, 99*:361, 1969.
19. Yip, B.S.S.C., and Jordan, P.H., Jr.: Radioimmunoassay of gastrin using antiserum to porcine gastrin. *Proc Soc Exp Biol Med, 134*:380, 1970.
20. Jeffcoate, S.L.: Radioimmunoassay of gastrin: Specifity of gastrin antisera. *Scand J Gastroenterol, 4*:457, 1969.
21. Ganguli, P.C., and Hunter, W.M.: Radioimmunoassay of gastrin. Abstracts of the IV World Congress of Gastroenterology, Copenhagen. July 1970.
22. Labo, G., Barbara, L., Vezzadini, P., and Corinaldesi, R.: Development of a radioimmunoassay for the measurement of gastrin. Abstracts of the IV World Congress of Gastroenterology, Copenhagen. July 1970.
23. Hansky, J., and Cain, M.D.: Radioimmunoassay of gastrin in human serum. *Lancet, 2*:1388, 1969.
24. Young, J.D., Byrnes, D.J., Chishold, D.J., Griffiths, F.B., and Lazarus, L.: Radioimmunoassay of gastrin in human serum using antiserum against pentagastrin. *J Nucl Med, 10*:746, 1969.

CHAPTER 2

THE CLINICAL SIGNIFICANCE OF GROWTH HORMONE AND INSULIN ASSAY

FRANCIS A. ZACHAREWICZ

INTRODUCTION

THE SUCCESSFUL DEVELOPMENT of the radioimmunoassay techniques ushered in a new and exciting era in medicine. The information and knowledge, so obtained, confirmed many concepts, disproved others and also unfolded new avenues and ideas. The immunoassay techniques for the measurement of insulin [1,2] were the first reported. These have been followed by the description of techniques for the assay of growth hormone, thyrotropin, ACTH, gonadotropins, parathormone, thyrocalcitonin, proinsulin and glucagon, to identify just a few.

This review is designed to present data on insulin and growth hormone assays and the relationship to clinical situations. No attempt will be made to describe the immunoassay techniques, since that would be beyond the scope of this article.

INSULIN

In normal-weight, nondiabetic individuals the overnight fasting plasma insulin levels are in the range of 5 to 25 μU/ml.[3,4] In the same normal individual in response to a glucose load the plasma insulin rises briskly, to maximum values between thirty and sixty minutes and returns to baseline values in three to four hours (Fig. 2–1). In marked contrast, the insulin response to a glucose load in a patient in maturity-onset type diabetes is illustrated in Figure 2–2. The values depicted in Figure 2–2 were ob-

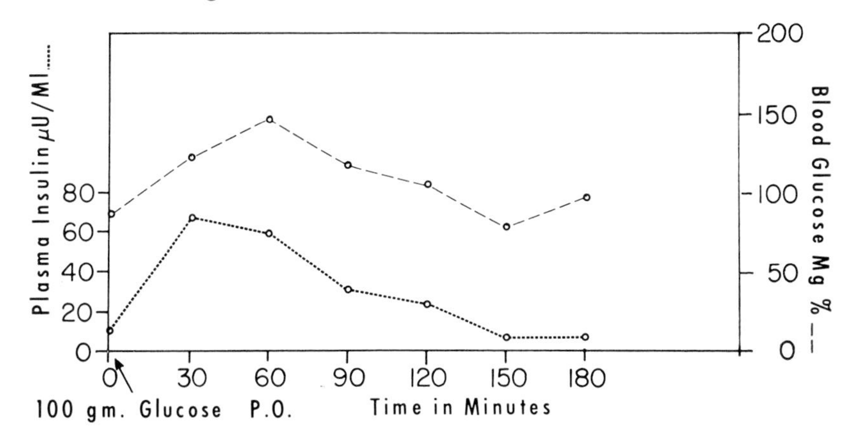

Figure 2–1. Insulin response curve to standard oral glucose load (100 gm) in a normal patient. The insulin response is brisk, maximum value is achieved in thirty minutes and returns to baseline in three hours.

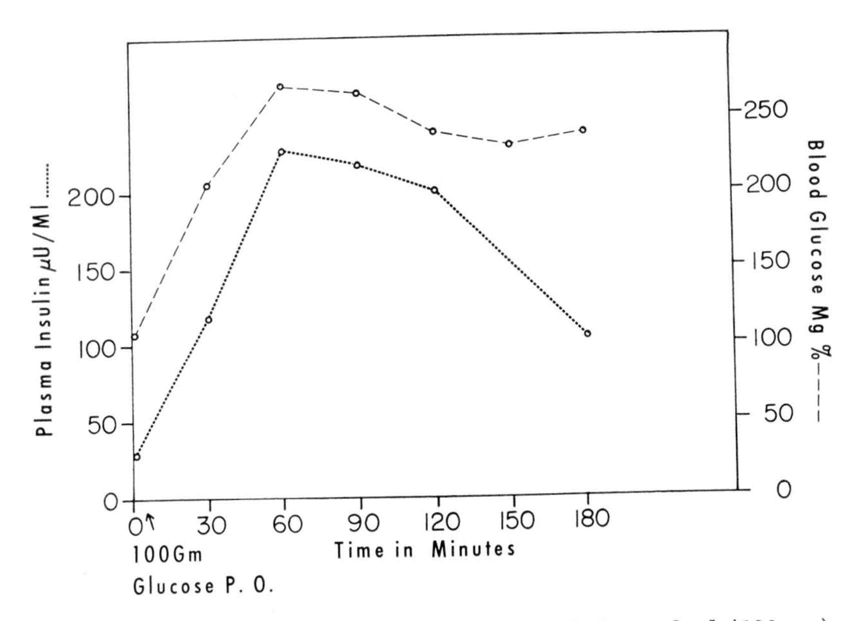

Figure 2–2. Insulin response curve to standard oral glucose load (100 gm) in a patient with maturity-onset diabetes. Fasting insulin level is mildly elevated; the rise of the insulin level is delayed, maximal response is higher than normal and fails to return to fasting level by 3 hours.

tained in a physician who had a positive family history for diabetes and was seen originally for symptoms of postprandial hypoglycemia. The major differences to be noted: The maximal response is generally higher than normal, the rise of the plasma insulin level is delayed, and the total amount of insulin secreted may be greater than normal.[3] The latter observation has been the subject of considerable interest and raises the question of the hypersecretion of insulin in the maturity-onset type diabetes. Seltzer et al.[5] compared the insulin level to the corresponding plasma glucose level in normals and diabetics. In the diabetic patients the ratio of plasma insulin to plasma blood glucose was less than that obtained in normal patients and therefore insulin secretion in maturity-onset diabetes is impaired. However, true hypersecretion of insulin without attendant hypoglycemia is frequently found in the third trimester of pregnancy,[6] obesity,[5,7] Cushing's syndrome and acromegaly.[8] In these pathologic states, the in-

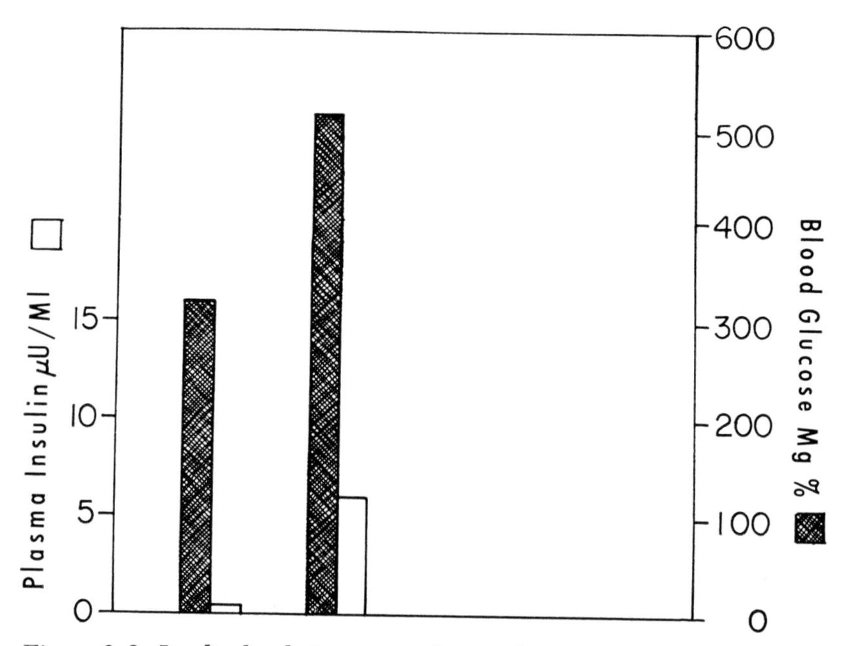

Figure 2–3. Insulin levels in untreated juvenile-type diabetes. Circulating insulin is undetectable or markedly depressed as depicted by the level of 6 μU per ml in a patient with a blood glucose of 515 mg %.

creased secretion of insulin is necessary to compensate for various anti-insulin factors or inhibitors which diminish the physiologic effect of insulin. Without this hypersecretory state, the glucose tolerance curve would fall in the diabetic range.

In juvenile or growth-onset type of diabetes, there is an absolute insulin deficiency. Figure 2–3 illustrates insulin levels obtained in two juvenile-type diabetics who have never been treated with either insulin or oral hypoglycemic agents. Despite the marked elevation of plasma glucose levels, one patient did not have any demonstrable insulin, while the second patient had only 6 μU/ml despite a plasma glucose of 515 mg percent. Although glucose tolerance tests were not performed on these patients, others have reported [9] the failure of oral glucose, and intravenous tolbutamide to stimulate insulin response in juvenile-type diabetes.

Factors That Regulate Normal Insulin Secretion

Prior to the extensive studies on insulin secretion made possible by the radioimmunoassay technique for measuring insulin, glucose was recognized as the only physiologic stimulant releasing insulin from the pancreatic beta cell. This view is no

TABLE 2–I
FACTORS STIMULATING INSULIN RELEASE

Glucose
Amino acids—L-leucine and Arginine
Gastrin
Secretion
Pancreozymin
Gut glucagon
Glucagon from a cell
Tolbutamide
B-adrenergic stimuli

longer accurate. It has now been determined that an interplay of nutrients, gut factors, nervous stimuli and other hormones alter insulin release from the beta cell. Table 2–I lists the various agents capable of stimulating insulin release. It is of interest to mention that epinephrine and norepinephrine are potent inhibitors of insulin release. The amino acids which stimulate insulin secretion have a synergistic effect with glucose on stimulating insulin release from the pancreas. In addition, both amino acids and glucose, when taken orally, stimulate a more prolonged and greater insulin response than when given intravenously.[10,11] This enhanced effect on insulin release is due to the various gut factors (glucagon, gastrin, secretin and pancreozymin) released from the gastrointestinal tract.

Usefulness of Insulin Assay

The ready availability of the radioimmunoassay technique for insulin determination expanded our fund of knowledge regarding both the physiologic and biologic mechanisms operative in insulin-glucose homeostasis. This has allowed a better understanding of glucose-insulin relationships in health and disease. As a rule, diabetes mellitus can easily be diagnosed with an oral glucose tolerance and insulin levels are not required. One area in which insulin levels have been most useful has been in the diagnosis of insulinoma and other causes of hypoglycemia.

Insulin Responses in Insulinoma

An insulinoma can present with signs and symptoms of hypoglycemia either during the fasting state or following stimulation with food ingestion. The beta cell tumor secretes insulin in an autonomous fashion, in excessive amounts which may result in hypoglycemia and the resultant clinical picture. Classically, Whipple's triad was used to diagnose this condition. The triad consists of a history of attacks of hunger, tremulousness, weakness, sweating and paresthesia occuring during the fasting period; a blood glucose level less than 40 mg percent during the attack, and prompt recovery with the administration of glucose. However, frequently it is necessary to resort to laboratory tests to document the presence of a pancreatic beta cell tumor. Table 2–II lists a number of diagnostic procedures, with the exception of the prolonged fast, that are provocative tests and depend on their ability to stimulate insulin release. It is unclear why some of the tests yield positive results while others are equivocal. Samols[10] repeated an intravenous tolbutamide test on the same patient five days apart. The first test yielded inconclusive information whereas the second test, performed five days later, was diagnostic of insulinoma.

TABLE 2–II
PROVOCATIVE STIMULATORS OF INSULIN RELEASE IN INSULINOMA

Oral Glucose Tolerance Test (100 g glucose)
Intravenous Tolbutamide (1 gram)
Oral L-Leucine Tolerance Test (150 mg/kg body wt)
Intravenous Glucagon Test (1 mg)

Oral Glucose Tolerance Test. On the whole, the oral glucose tolerance test is inconsistent in the results obtained in patients with insulinoma. The glucose responses may be flat, diabetic or a delayed type of hypoglycemia may be seen. In addition, the insulin responses to the glucose may be variable. Figure 2–4 depicts an oral glucose tolerance test following the ingestion of 100 grams of glucose. The plasma glucose levels are generally stable; the insulin levels, however, are excessive and greater than that expected with functional hypoglycemia.

Intravenous Tolbutamide Tolerance Test. The basis for the diagnostic value of intravenous tolbutamide in insulinomas rests on the fact that the majority of insulinomas are exquisitely

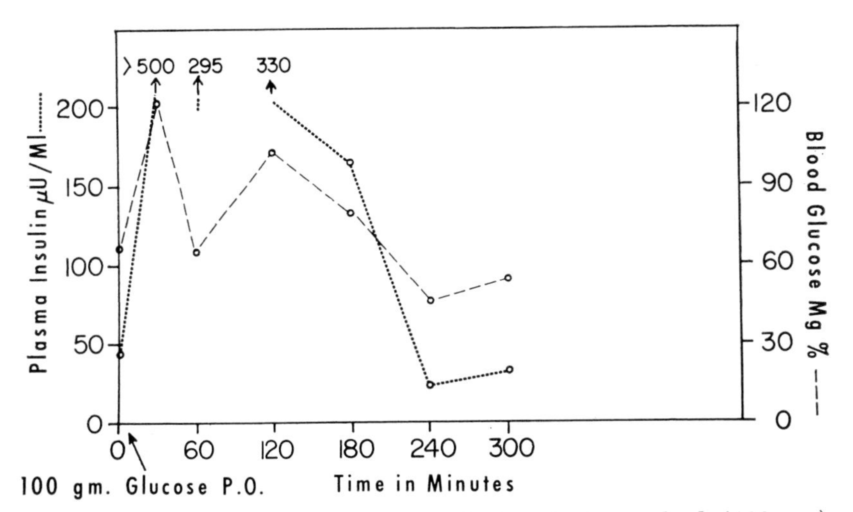

Figure 2–4. Insulin response to a standard oral glucose load (100 gm) in a patient with proven insulinoma. Fasting level of insulin is elevated and response to glucose load is markedly excessive as shown by the level of circulating insulin greater than 500 μU per ml in the presence of a blood glucose level of 120 mg percent.

sensitive to its stimulatory effect on the beta cell. Samols[10] considers that a rise in plasma insulin concentration greater than 100 μU/ml over fasting values ten to twenty minutes after intravenous tolbutamide as excessive. Figure 2–5 shows the response to one gram of intravenous tolbutamide in a patient with an islet cell tumor. In addition to the marked elevation of the plasma insulin levels, the blood glucose levels fell and remained low throughout the procedure.

ORAL L-LEUCINE TOLERANCE TEST. Cochrane initially identified L-leucine as the agent inducing hypoglycemia in infants with idiopathic infantile hypoglycemia.[11,12] Subsequently, Flanagan et

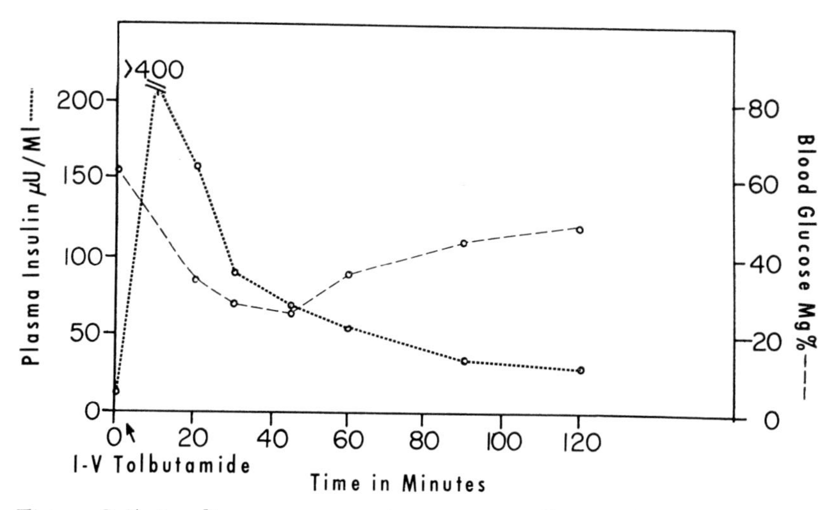

Figure 2–5. Insulin response to intravenous tolbutamide administration (1 gram) in a patient with an islet cell tumor. The responsive tumor tissue secretes excessive quantities of insulin which lowers blood glucose levels to hypoglycemic levels.

al.[13] reported the hypoglycemic effect of leucine in patients with functioning islet cell tumors. A 40 percent or greater fall in blood glucose levels is considered diagnostic of insulinoma, provided the patient has not been pretreated with an oral sulfonylurea. The leucine solution is prepared as follows: 150 mg per kg of body weight of L-leucine is dissolved in water as a 2 percent solution and ingested over a five-minute period. After a control blood sample is taken, the leucine solution is ingested and blood specimens are obtained at 10, 15, 30, 60, 90 and 180 minutes after leucine. Figure 2–6 depicts the results obtained in a patient with an insulinoma. Note the 60 percent fall in blood glucose and the excessive rise in plasma insulin levels. In normal individuals, the blood glucose values remain stable and unchanged from the fasting level, whereas the plasma insulin levels will only increase two- to three-fold, which is less than that seen in patients with functioning islet cell tumors.

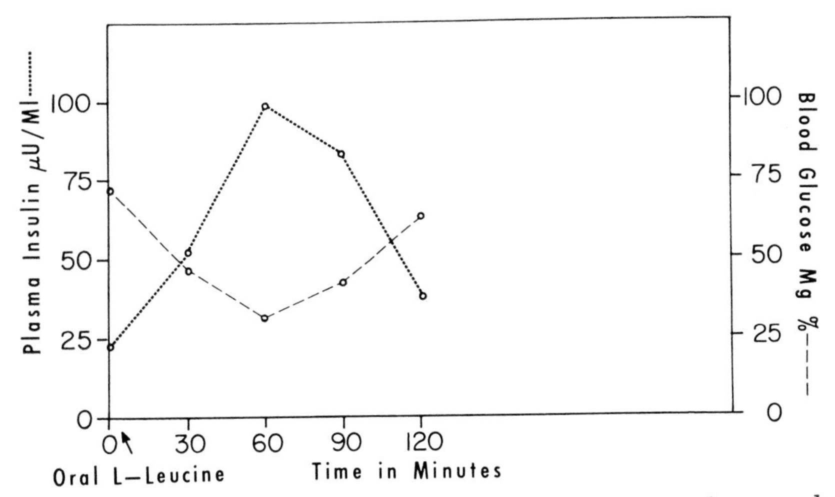

Figure 2–6. Effect of oral L-leucine administration on blood glucose and plasma insulin in a patient with an insulinoma. The blood glucose decrease is greater than 40 percent of the fasting level and the insulin level exceeds the normal two-fold increase over the basal level.

GLUCAGON STIMULATION TEST. Over and above glucagon's effect as a glycogenolytic agent in the liver, glucagon directly stimulates insulin release from pancreatic islet tissue. In patients with islet cell tumor, this response is greatly exaggerated,[14] and values of insulin greater than 180 µU/ml are considered diagnostic. The test is performed by obtaining a control blood sample, then injecting 1 mg of glucagon intravenously and specimens drawn at 10, 20, 30, 45, 60, 90 and 120 minutes after injection. Maximum rise in plasma insulin levels is seen within the first thirty minutes. Blood glucose levels fall after an initial rise (Fig. 2–7).

Effect of Prolonged Fasting. The most common abnormality in patients with functioning islet cell tumors is the inappropriate insulin release during the prolonged fast. A majority of patients will be symptomatic with blood glucose levels less than 40 mg percent during the first twenty-four hours of the fast. At the same time the insulin level will be inappropriately elevated for

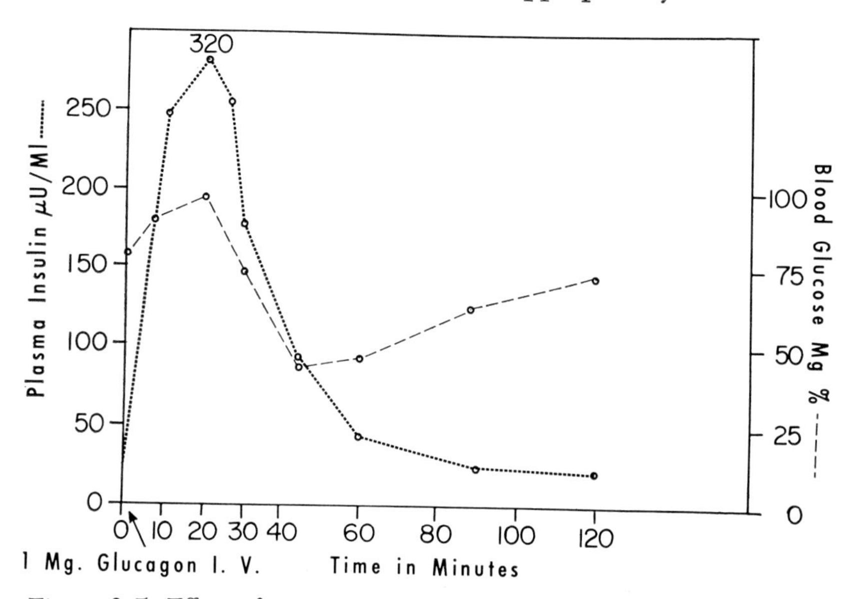

Figure 2–7. Effect of intravenous glucagon administration on blood glucose and plasma insulin in a patient with an insulinoma. The insulin response of 320 μU per ml is excessive when compared to accepted maximal increase to 180 μU per ml in normal individuals. The rise in circulating insulin level is generally associated with marked lowering of the blood glucose level as shown in this patient.

the level of blood glucose.[15,16] Patients are on a total fast except for water *ad lib.* If a positive result (symptomatic hypoglycemia) is not experienced during the fasting period, it is followed by a short period of exercise, as climbing stairs or pedaling a stationary bicycle. Blood specimens for glucose and insulin determinations are obtained every four to six hours throughout the test and immediately whenever the patient becomes symptomatic.

This review has dealt primarily with hypoglycemia associated with insulinoma. Needless to say, there are numerous causes of both spontaneous and stimulative-type hypoglycemia (alimentary hyperglycemia, functional hypoglycemia), but the measurement of plasma insulin levels represents the most useful application of the immunoassay at present—especially following the test procedures as outlined in this review. It should be pointed out that, as in any other laboratory procedures, norms and basal insulin values should be established in each hospital.

GROWTH HORMONE

Prior to the development of the sensitive immunological assay for plasma human growth hormone (HGH),[17] the physician was dependent solely on nonspecific and insensitive techniques for diagnosing acromegaly. Those included determination of serum phosphorus concentrations, glucose tolerance test, basal metabolic rate, insulin tolerance test and x-rays. These tests are insensitive and unreliable in diagnosing acromegaly in its early stage, prior to the development of irreversible anatomic changes.

Direct immunoassay of HGH has solved this problem of early diagnosis. In addition, considerable information has been obtained regarding the physiology of growth hormone secretion. It is now known that growth hormone secretion continues through adult life.[17] After overnight fast, normal adult values for growth hormone are usually between 3 and 10 mμg/ml. Table 2–III lists the various factors that stimulate a rise in growth hormone levels.[18] With the exception of major surgery, pyrogen, ingestion of a high protein meal and intravenous amino acids, this rise in plasma growth hormone level is related to a carbohydrate deficiency. Direct glucose administration produces a rapid decline in plasma levels, usually to unmeasurable levels. The effects of glucose administration and hypoglycemia on plasma growth hormone levels are significant because they establish the normal control of growth hormone secretion. Roth et al.[18] have reported a normal patient who had 50 mμg/ml of plasma HGH, but suppressed following glucose administration to values less than 4 mμg/ml.

In normal individuals following an oral glucose tolerance test, plasma HGH levels fall to values below 5 mμg/ml. Failure to fulfill these criteria yields unequivocal evidence of hypersomatotropism. The administration of intravenous insulin 0.1 to 0.2 μ/kg body weight to decrease fasting blood glucose levels by 40 percent or more stimulates HGH release.[19] Figures 2–8A and 2–8B demonstrate this effect of glucose and insulin on plasma growth hormone levels. Arginine stimulates a rise in plasma HGH levels similar to that obtained with intravenous insulin but without the potential hazards of the induced hypoglycemia.[20] Figure 2–9 shows the pattern of HGH response obtained with glucose and intravenous insulin in a patient with acromegaly. These results

TABLE 2–III
FACTORS STIMULATING HUMAN GROWTH HORMONE SECRETION

Factors
Prolonged fasting
Hypoglycemia
Rapid fall in blood glucose
Psychic stress
Ingestion of a high protein meal
Muscular exercise
Inhibition of intracellular glucose utilization
Pyrogen
Intravenous amino acids
Major surgery
Onset of sleep

indicate that a glucose meal fails to shut off HGH secretion, and hypoglycemia, a normal stimulant of HGH secretion, is ineffective in raising the plasma level. Since the basal level is elevated and is neither suppressible nor stimulated, it represents inappropriate HGH secretion consistent with acromegaly. This diagnosis can be made early before significant irreversible anatomic changes have developed.

The value of determining HGH levels following therapy of acromegaly as a criteria of successful therapy is unsettled. Conventional x-ray and gamma ray therapy, heavy particle therapy, yttrium 90 implantation, direct surgical hypophysectomy and cryogenic surgery have been used to treat acromegaly. Of these, conventional pituitary irradiation and heavy particle therapy has been studied extensively. Roth et al.[21] have reported on the success of conventional pituitary irradiation in acromegaly and

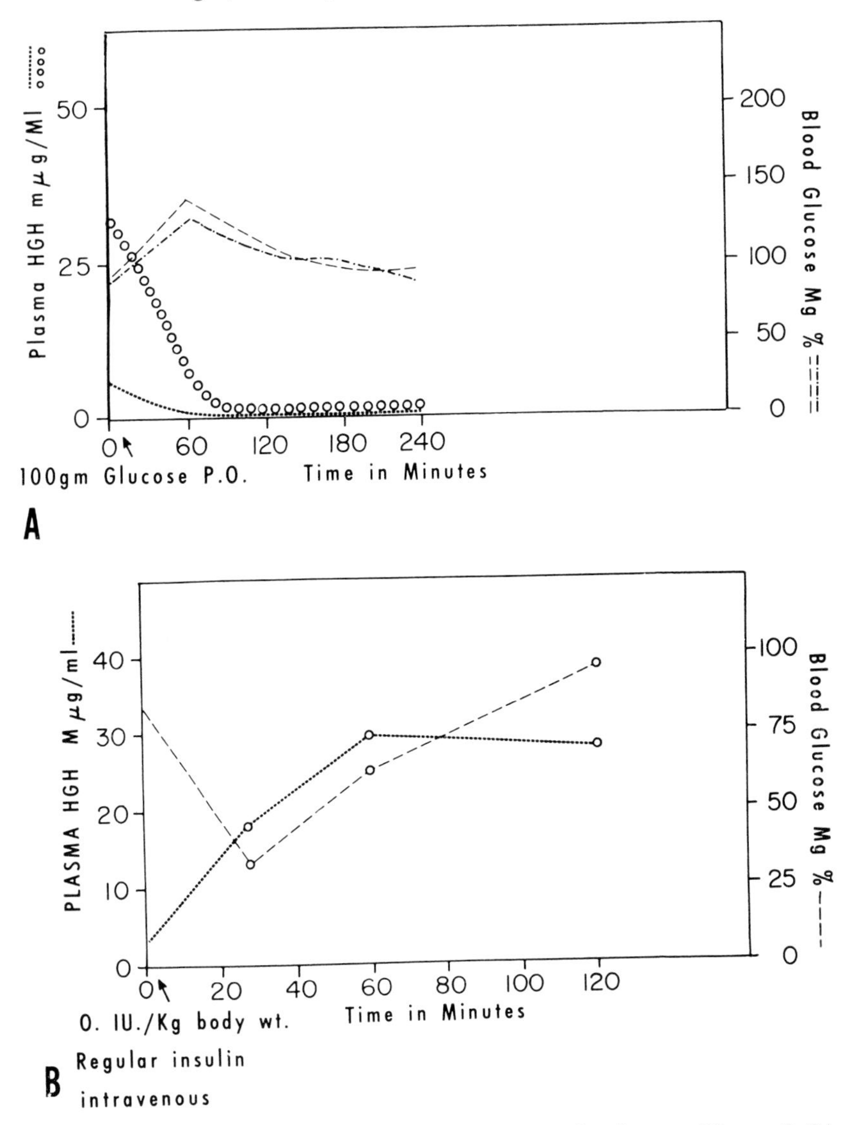

Figure 2–8. Growth hormone response in normal subjects. Figure 2–8A illustrates the suppression of circulating growth hormone to values less than 5 mμg per ml to a standard oral glucose load (100 grams). Figure 2–8B shows the normal increase of growth hormone following insulin-induced hypoglycemia. These results illustrating suppression and stimulation of circulating growth hormone level demonstrate normalcy of growth hormone secretion.

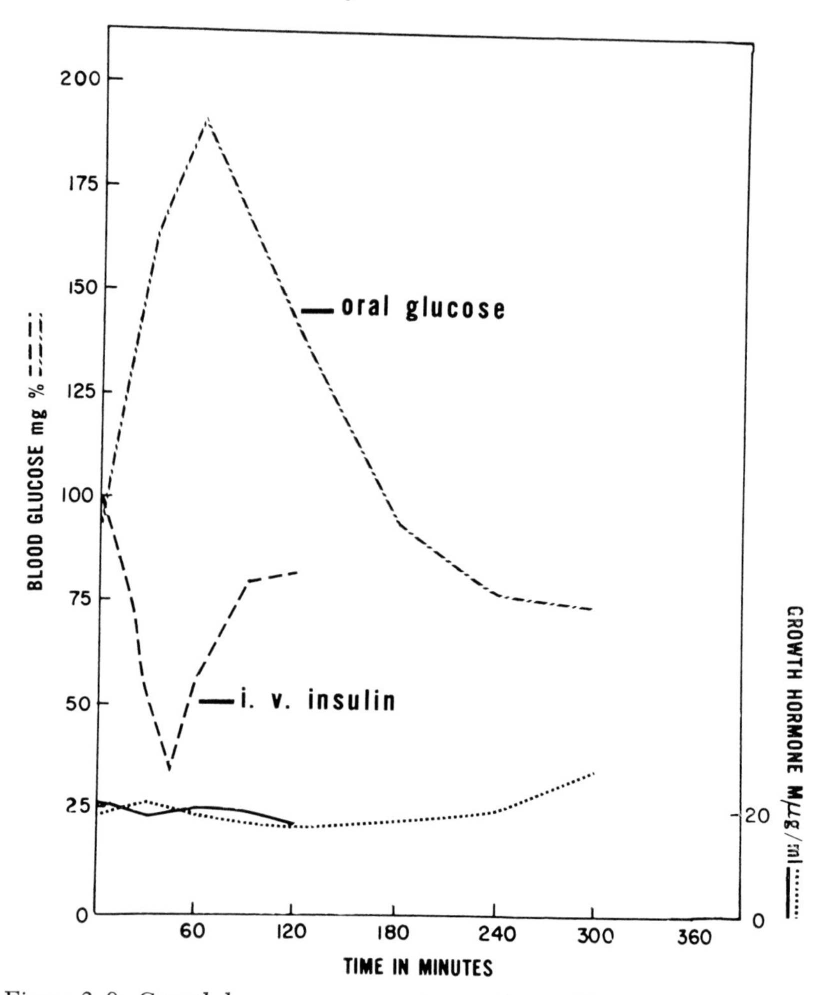

Figure 2–9. Growth hormone response in a patient with active acromegaly. The administration of either a standard oral glucose load (100 grams) or induction of insulin-induced hypoglycemia fail to suppress or increase the circulating level of growth hormone. This response is indicative of autonomous secretion of growth hormone and is diagnostic of acromegaly.

supported their conclusions with the observed fall in plasma HGH and clinical improvement. Linfoot et al.[22] have reported on the superiority of heavy particle therapy as compared to conventional pituitary irradiation. Both methods of therapy lower HGH levels and demonstrate some degree of suppressibility to a glucose meal. This aspect of acromegaly requires additional investigation.

The major contribution of the radioimmunoassay of human growth hormone resides in the ability of the technique to detect excessive HGH production in its early phases and prior to the development of anatomic changes in acromegaly.

REFERENCES

1. Berson, S.A., and Yalow, R.S.: Immunoassay of endogenous plasma insulin in man. *J Clin Invest, 39:*1157, 1960.
2. Grodsky, G.M., and Forsham, P.H.: An immunochemical assay of total extractable insulin in man. *J Clin Invest, 39:*1070, 1960.
3. Yalow, R.S., and Berson, S.A.: Plasma insulin concentrations in nondiabetic and early diabetic subjects. Determinations by a new sensitive immunoassay technique. *Diabetes, 9:*254, 1960.
4. Perley, M., and Kipnis, D.M.: Effect of glucocorticoids on plasma insulin. *N Engl J Med, 274:*1237, 1966.
5. Seltzer, H., Allen, E.W., Herron, A., Jr., and Brennan, M.: Insulin secretion in response to glycemic stimulus: Relation of delayed initial release to carbohydrate intolerance in mild diabetes mellitus. *J Clin Invest, 46:*323, 1967.
6. Kalkhoff, R., Schalch, D.S., Walker, J.L., Beck, P., Kipnis, D.M., and Daughaday, W.H.: Diabetogenic factors associated with pregnancy. *Trans Assoc Am Phys, 77:*270, 1964.
7. Perley, M., and Kipnis, D.M.: Plasma insulin responses to glucose and tolbutamide of normal weight and obese diabetic and nondiabetic subjects. *Diabetes, 15:*867, 1966.
8. Beck, P., Schalch, D.S., Parker, M.L., Kipnis, D.M., and Daughaday, W.H.: Correlative studies of growth hormone and insulin plasma concentrations with metabolic abnormalities in acromegaly. *J Lab Clin Med, 66:*366, 1965.
9. Chiumello, G., Del Guericio, M.J., and Bidone, G.: Effects of glucagon and tolbutamide on plasma insulin levels in children with ketoacidosis. *Diabetes, 17:*133, 1968.
10. Samols, E., and Marks, V.: Evaluation of the intravenous tolbutamide test in the recognition and differential diagnosis of spontaneous hypoglycemia. In Butterfield, W.J.H., and Van Westering, W. (Eds.): *Tolbutamide after Ten Years.* Proceedings of the Brook Lodge Symposium, Augusta, Michigan, 1967. International Congress Series No. 149. New York, Excerpta Medica Foundation, 1967, pp. 147–156.
11. Cochrane, W.A., Payne, W.W., Simkiss, M.J., and Woolf, L.I.: Fa-

milial hypoglycemia precipitated by amino acids. *J Clin Invest, 35*:411, 1956.

12. Cochrane, W.A.: Studies in the relationship of amino acids to infantile hypoglycemia. *AMAS Dis Child, 99*:476, 1960.
13. Flanagan, G.C., Schwartz, T.B., and Ryan, W.G.: Studies on patients with islet-cell tumors, including the phenomenon of leucine-individual accentuation of hypoglycemia. *J Clin Endocrinol, 21*:401, 1961.
14. Fajans, S.S.: Diagnostic tests for functioning pancreatic islet-cell tumors. *Excerpta Medica, 172*:894, 1967.
15. Yalow, R.S., and Berson, S.A.: Dynamics of insulin secretion in hypoglycemia. *Diabetes, 14*:341, 1965.
16. Scholz, D.A., ReMine, W.H., and Priestley, J.T.: Clinics on Endocrine Metabolic Diseases, 3. Hyperinsulinism: Review of 95 cases of Functioning Pancreatic islet-cell tumors. *Staff Meeting, Mayo Clinic, 35*:545, 1960.
17. Glick, S.M., Roth, J., Yalow, R.S., and Berson, S.A.: Immunoassay of human growth hormone in plasma. *Nature (Lond), 199*:784, 1963.
18. Roth, J., Glick, S.M., Yalow, R.S., and Berson, S.A.: Secretion of Human Growth Hormone: physiologic and experimental modification. *Metabolism, 12*:577, 1963.
19. Frantz, A.G., and Rabkir, M.T.: Human Growth Hormone—Clinical measurement, response to hypoglycemia and suppression by corticosteroids. *N Engl J Med, 271*:1375, 1964.
20. Merimee, T.J., Rabinowitz, D., Riggs, L., Burgess, J.A., Rimon, D.L., and McKusick, V.A.: Plasma Growth Hormone after arginine infusion. *N Engl J Med, 276*:434, 1967.
21. Roth, J., Gorden, P., and Brace, K.: Efficacy of conventional pituitary irradiation in acromegaly. *N Engl J Med, 282*:1385, 1970.
22. Linfoot, J.A., Garcia, J.F., Hoye, S.A., Schmitt, J., and Lawrence, J.H.: Heavy particle therapy in acromegaly. *Proc R Soc Med, 63*:219, 1970.

CHAPTER 3

RADIOIMMUNOASSAY FOR PROSTAGLANDINS

Bernard M. Jaffe and Charles W. Parker *

INTRODUCTION

ALTHOUGH A MYRIAD of activities have been ascribed to the prostaglandins, their physiologic significance has not yet been evaluated. Forty years have elapsed since two gynecologists, Kurzrok and Lieb, reported that human seminal fluid stimulated strips of uterus from women who had been sterile and relaxed strips of uterus from women who had undergone successful pregnancies.[1] These experiments gave the first indication that semen contained some constituent(s) which had a differentiated biological activity. In 1935 Goldblatt reported that extracts of seminal fluid stimulated smooth muscle and lowered blood pressure.[2] The name prostaglandin was coined by von Euler in 1935 [3] because the active lipid in semen was thought to originate from the prostate and vesicular gland. Several years elapsed from then until 1960 when Bergström and Sjövall successfully isolated an active compound from sheep vesicular gland [4] and 1962 when this group of investigators described the chemical structure of the first of the prostaglandins.[5]

Chemically, the prostaglandins are twenty carbon polyenoic fatty acids. Formed from unsaturated fatty acid precursors, the basic chemical structure consists of a cyclopentane ring and two linear side chains. Four basic series of prostaglandins have been

* This work was supported in part by USPHS Training Grant GM 371. We wish to gratefully acknowledge the cooperation of the Upjohn Company (Dr. John Pike) for supplying the prostaglandins, the Ayerst Company (Dr. R.O. Davis) for supplying the prostaglandin analogs, Drs. F. Matschinky and M.A. Marazzi for supplying the enzymatically prepared PGE_1, Dr. J.W. Smith for his capable collaboration, and Mrs. D. Collier for her skilled technical assistance.

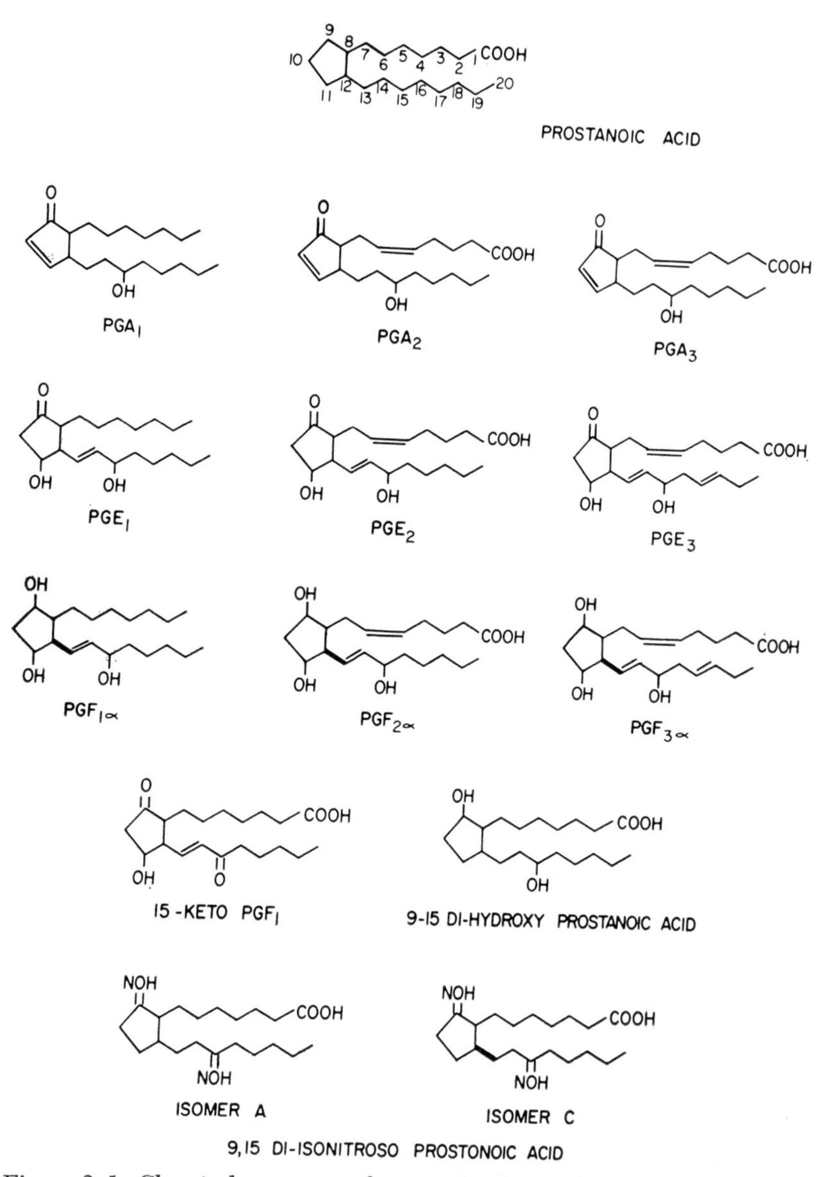

Figure 3–1. Chemical structure of prostaglandins and related analogs.

described (A, B, E and F), the division being based on differences in substitutions on the cyclopentane ring (Fig. 3–1). Each series is further subdivided by the number of unsaturated bonds in the side chains corresponding to the subscript following the series designation. Prostaglandins all have at least one double bond, at the 13–14 position; the second double bond occurs at the 5–6 position and the third at the 17–18 position. These double bonds can be either of the cis or trans variety. Almost all active prostaglandins have a hydroxyl group at the 15 position.[6,7]

The multiple pharmacologic effects of the various prostaglandins described have been correlated with the specific prostaglandin as well as the species, tissue and conditions tested. Basically, prostaglandin (PG)E has both vasodepressor and smooth muscle activities, PGF has primarily smooth muscle effects and PGA is almost purely a vasodepressor. As an index of the rapid proliferation of reports on prostaglandins, approximately two thirds of the reports published by 1970 were published in 1968 and 1969. This profusion of pharmacologic data has not, however, clarified the physiologic role of the prostaglandins. The most serious obstacle to the elucidation of the function of this family of compounds is the lack of a suitable method for their measurement. Bioassay systems test predominantly effects on either smooth muscle or on blood pressure. Assays can be performed on rabbit and guinea pig intestine,[8-10] rat stomach fundus[11] and uterus[10,13] and hamster colon.[12,14] Vasodepressor activity in rabbits, rats and cats can also be used for bioassay.[15-18] These bioassay systems are sensitive primarily in the nanogram range.

Existing methods for chemical determination of prostaglandins are useful in the high nanogram or microgram range. They commonly utilize a base-catalyzed dehydration rearrangement which converts PGE to PGA and thence to PGB. PGA has an absorption maximum at 217 nm, PGB at 278 nm. Converted to PGB, the molar extinction coefficient (ethanol, 278 nm) for PGE and PGA is $27{,}200 \pm 500$.[19] Gas chromatography and mass spectometry have also been applied to prostaglandin measurement.[20]

The predominant pathway for prostaglandin degredation is

by oxidation of the hydroxyl at the 15 position to a keto derivative. The responsible enzyme, prostaglandin —15 dehydrogenase, has been isolated from swine lung.[21-23] Enzymatically prepared 15 keto prostaglandin can be estimated either by measuring the chromophore at 500 nm which forms on incubation of the reaction mixture with sodium hydroxide or by extracting and purifying the 15 keto compounds from the reaction mixture and measuring them directly by their absorption at 230 nm.[19]

None of these types of assays—biological, chemical, or enzymatic—is sensitive enough to measure concentrations of prostaglandins in physiologic situations. In this report we describe a specific radioimmunoassay for prostaglandins capable of quantitating subpicomolar amounts of prostaglandins.[23] Application of this radioimmunoassay system to the study of prostaglandins may finally elucidate their functional role.

PREPARATION OF RADIOACTIVE-LABELED PROSTAGLANDIN

Critical to the success of a functioning radioimmunoassay is the development of a satisfactory radioactive label.

The initial radioactive prostaglandin utilized was prepared according to the method of Samuelsson[24] which basically consisted of catalytic reduction of PGE_2 to PGE_1, using tritium gas. PGE_2, 12.7 mg dissolved in 1.5 ml ethyl acetate and 45 mg palladium black 5% on charcoal were frozen in liquid nitrogen in a vial with a siliconized rubber stopper. After evaporation of solvent with high vacuum, 5 curies of tritium gas were introduced. The reaction mixture was allowed to warm to room temperature, was stirred for twenty minutes and then refrozen. The product was partially purified by thin layer chromotography using silica gel impregnated with 10 percent silver nitrate. The solvent system was chloroform-tetrahydrofuran-glacial acetic acid 10:2:1.[25] The specific activity of this product was less than 1 Ci/mM, and hence will be referred to as low specific activity PGE_1.

High specific activity (^3H) PGE_1, 87 Ci/mM, was purchased from the New England Nuclear Corporation. It was used in all but the initial antibody studies, and is currently utilized in the radioimmunoassay system to be described later. Additional pure high specific activity prostaglandins are available from that source including PGA_1 (30 to 80 Ci/mM), PGE_2 (6 Ci/mM), PGF_1 (40 to 60 Ci/mM) and PGE_2 (5 to 10 Ci/mM).

Attempts to produce iodinated labels have largely failed. Direct iodination of prostaglandins and iodination of tyrosine-prostaglandin conjugates disrupts the ring structure sufficiently to prevent antibody binding. An alternative method has previously been employed in which the hormone-ligand is bound to a tryosine containing copolymer. When this complex is iodinated it functions as a labeled hormone; this approach has been used for gastrin tetrapeptide [26,27] bradykinin [28] and estrogens.[29] This same technique has been used with limited success for prostaglandin both by the author and Levine and van Vanukis,[30] both groups of which utilized random copolymers of Glu-Ala-Lys-Tyr, mole ratios 36,35,24,5 respectively. The author used both carbodiimide and ethyl chloroformate to conjugate PGA_1 to the copolymer using trace amounts of (^3H) PGA_1; to follow the reaction Levine and van Vanukis used carbodiimide and PGE_1. The conjugates were iodinated by the method of Hunter and Greenwood.[31] Both types of (^{125}I) copolymer-prostaglandin were specifically bound by antiprostaglandin antibody and this binding was inhibited by added unlabeled prostaglandin. However, sensitivity of radioimmunoassay with these iodinated labels was severely limited (to the nanogram range).

PRODUCTION OF ANTIPROSTAGLANDIN ANTIBODIES

Antibodies to a low molecular weight fatty acid might have been difficult to elicit. The immunogens we utilized for stimulating antibodies were prostaglandin-protein conjugates. In the conjugation reaction, advantage was taken of the free carboxylic acid group on prostaglandin which was covalently bonded to amino groups on the carrier protein. In addition, serum albumin was chosen as the predominant carrier because of its known affinity for prostaglandin. In equilibrium dialysis chambers, labeled prostaglandin concentrated on the albumin-containing side of the membrane (HSA 1 and 10 mg/ml); applying the Scatchard equation (32) $\frac{r}{c} = nka - rka$, to the data, the association constant of albumin for PGE_1 is at least 10^5 L/M.

The conjugates prepared are summarized in Table 3–I. Protein conjugates of PGA_1, PGA_2, PGE_1, PGE_2 and PGF_1 were prepared using two different carbodiimides.[33,34] In a typical reaction,

TABLE 3–I
IMMUNOGENS UTILIZED IN PRODUCING ANTIPROSTAGLANDIN ANTIBODIES

Prostaglandin	*Protein Carrier*	*Conjugating Agent*	*Degree of Substitution* (moles PG / mole carrier)
PGA_1	Human serum albumin	EDC *	3.3
PGA_2	Human serum albumin	EDC	2.7
PGE_1	Human serum albumin	EDC	1.5
PGE_1	Keyhole limpet hemocyanin	Ethylchloroformate	—
PGE_2	Bovine serum albumin	CMC †	3.8
PGF_1	Bovine gamma globulin	CMC	—

* 1-ethyl-3-(3-diethylaminopropyl) carbodiimide hydrochloride
† 1-cyclohexyl-3-morphinolyl-(4) ethyl carbodiimide metho-p-toluenesufonate

4 mg (12 micromoles) of PGA_1, dissolved in 4 ml 10 percent ethanol-20 percent aqueous sodium carbonate were incubated for twenty-four hours at 20° C and at pH 5.5 with 8 mg of human serum albumin (0.1 micromole) and 4 mg of l-ethyl-3-(3 diethylaminopropyl) carbodiimide hydrochloride (20 micromoles). The product was dialyzed over several days versus several liters of 0.15 M NaCl-0.01 M phosphate, pH 7.5. Reaction products were lyophilized.

Using an alternative conjugating agent, ethyl chloroformate,[35,36] PGE_1 was bound to keyhole limpet hemocyanin. Two milligrams of PGE_1 was reacted at 4° C for 15 minutes with equimolar quantities of ethyl chloroformate, and triethylamine following which 5.5 mg of hemocyanin was added in 0.45 ml of O.1M $NaHCO_3$. The reaction was allowed to proceed for one hour and then the reaction mixture was extensively dialyzed versus PBS.

The degree of substitution of prostaglandin on the protein

carriers was determined by observing the increase in the absorbency at 278 nm as PGA_1 and PGE_1 in alkali are converted to PGB derivatives.[19] Based on their molar extinction coefficients in alkali (corrected for small effects on protein and carbodiimide absorbency), from 1.5 to 3.8 moles of prostaglandin was calculated to have been conjugated to each mole of protein.

Lyophilized conjugates were dissolved in phosphate buffered saline and emulsified with equal parts of Freund's adjuvant containing heat-killed *M. tuberculosis.*[37] Rabbits were immunized at from two- to four-month intervals. At the initial immunization one milligram was injected subcutaneously in the foot pads (0.25 mg in 0.1 ml per foot pad); at subsequent immunization, a total

TABLE 3–II
SPECIFICITY OF EARLY ANTIPROSTAGLANDIN ANTIBODIES

	Percent inhibition of bound (^3H) PGE_1 in the presence of 50 ng of		
Serum	*PGA_1*	*PGE_1*	*PGE_2*
Anti-PGA_1-HSA	54%	35%	20%
Anti-PGA_2-HSA	46%	0	8%
Anti-PGE_1-KLH	42%	40%	32%

of from 50 to 150 micrograms were injected. Ten days after the booster immunizations, the rabbits were bled by cardiac puncture and the sera stored at −70° C.

All animals immunized made antiprostaglandin antibodies. Compared to normal rabbit sera, gamma globulin fractions from early antibody-containing sera prepared by ammonium sulfate precipitation followed by batch elution from DEAE cellulose [38] bound up to 20 percent of the low specific activity (^3H) PGE_1. In the presence of carrier normal rabbit gamma globulin to insure complete precipitation of antibody globulin, antibody-bound (^3H) PGE_1 was separated from free label by addition of ammonium sulfate to a final concentration of 50 percent of saturation.

Binding of low specific activity (^3H) PGE_1 by early antibody was specifically inhibited by the addition of 20 to 20,000 ng of unlabeled PGE_1. Antibody specificity was directed against determinants on both the cyclopentane ring and the aliphatic side chains. Data in Table 3–II demonstrated that 50 ng of PGA_1,

PGE_1 and PGE_2 inhibited the binding of low specific activity (^3H) PGE_1 to varying degrees depending predominantly on the immunogen utilized to elicit the antibodies tested. In general, the homologous compounds were the most effective inhibitors of the binding of label.

Sera obtained seven to ten days following late immunization (12 to 16 months) have had considerably higher liters. Using pure (^3H) PGE_1 of high specific activity, 0.05 ml of a 1:250 dilution of either anti-PGE_1 and anti-PGA_1 antisera binds more than 90 percent of the tritiated label; 0.05 ml of a 1:10,000 dilution of the same antisera binds half of the radioactive marker.

The better binding afforded by the use of higher titer antibody and greater specific activity (^3H) PGE_1 has allowed for more definitive evaluation of antibody specificy. Data in Figure

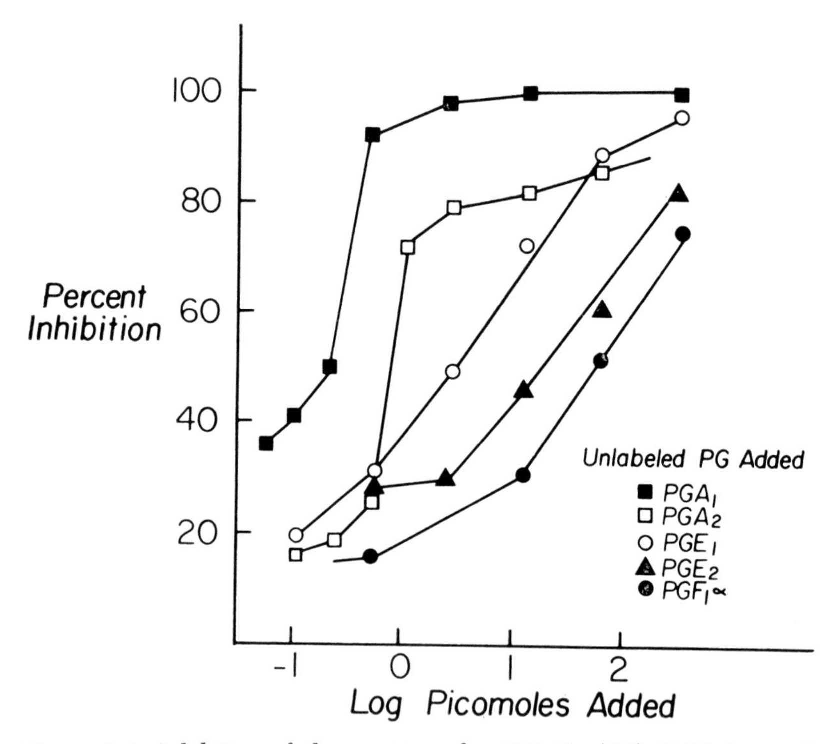

Figure 3–2. Inhibition of the reaction of anti-PGA_1–(^3H) PGE_1 by equimolar concentration of PGA_1, PGA_2, PGE_1, PGE_2 and PGF_{1a}.

3–2 confirmed the presence of determinants on both the cyclopentane ring and the alphatic side chains. In the experiment illustrated, anti-PGA_1 antibody was incubated with (^3H) PGE_1 in the presence of equimolar concentrations of PGE_1, PGE_2, PGA_1, PGA_2 and $PGF_1\alpha$. In the absence of unlabeled prostaglandin inhibitors 40 percent of the (^3H) PGE_1 was specifically bound by antibody. The percentage inhibition by prostaglandins of the expected binding of tritiated label plotted was calculated from the difference in binding in the presence and absence of the prostaglandins tested. PGA_1, the homologous ligand, inhibited significantly when only 0.1 picomole was included in the reaction mixture. The next most effective inhibitor, PGA_2, varied from the homologous only in the side chain. The fact that these two inhibition curves were relatively parallel and could be extrapolated to total inhibition suggests relatively homogeneous binding sites.[39] Members of the E series of prostaglandins were still less effective inhibitors, but as with the PGA's, the monounsaturated prostaglandin cross-reacted to a greater degree than did the corresponding molecule with two unsaturated double bonds. PGF_1 has a hydroxyl derivative in the 9 position rather than a keto group and as such inhibited rather poorly.

Similar data utilizing equimolar concentrations of prostaglandin analogs as inhibitors (with a different anti-PGA_1 antibody) further demonstrate the specificity of the antibodies. Compared to the homologous ligand, PGA_1, 9,15 dihydroxyprostanoic acid was only about one-third as potent as inhibitor (Fig. 3–3A). The enzymatically prepared 15 keto PGE_1, on the other hand, inhibited only a negligible amount even at the highest concentration tested. This means that the common degradation product of prostaglandins is so poorly cross-reactive that it will not interfere significantly with the measurement of biologically active prostaglandins when they are both present in a sample.

Illustrative of the extreme specificity of the antibody specificity is the fact that "Isomer A" and "Isomer C," which have significant inhibition differences (as seen in Fig. 3–3B), differ from each other chemically only as diastereoisomers at the 12 position of 9, 15 diisonitrosoprostanoic acid.

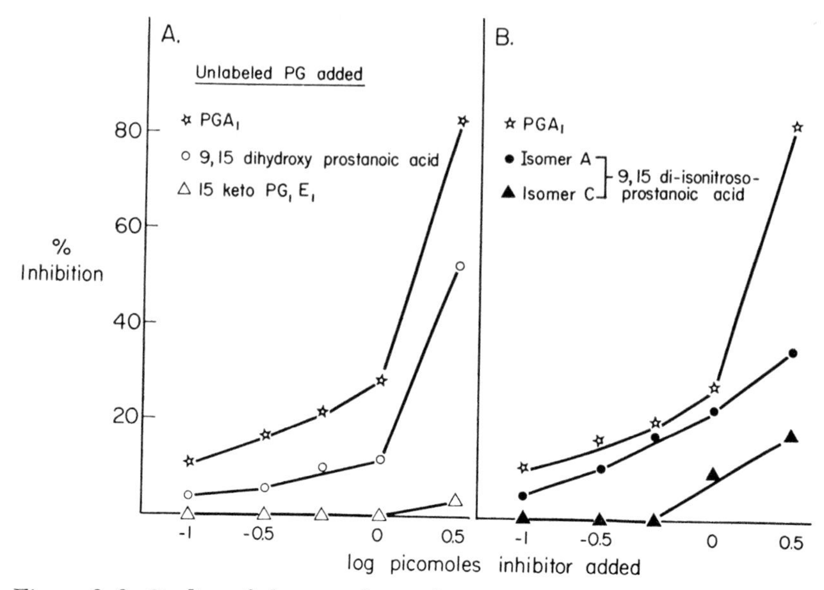

Figure 3–3. Studies of the specificity of an anti-PGA_1 antibody by demonstrating the degree of inhibition of (^3H) PGE_1 binding by several prostaglandin analogs including (A) 9, 15 dihydroxy prostanoic acid and 15 keto PGE_1 and (B) two diastereoisomers of 9, 15 di-isonitroso prostanoic acid. The chemical structures of all these compounds are illustrated in Figure 3–1.

In addition to studying the antibody specificity with reference to a variety of prostaglandins, we have also examined the cross-reactivity with a number of pharmacologic agents and lipids. Since most of these compounds circulate in plasma, it would be important to exclude their interference in a functioning radioimmunoassay system. Used at 10^{-5}M concentrations, these compounds were incubated with anti-PGA_1 antibody in the presence of tritiated prostaglandin. None of the fat soluble vitamins evaluated (A, D_2, E, K or K_3) cross-reacted significantly. Likewise, compounds which affect cyclic-AMP or the adenyl cyclase system,[40] including theophyllin, norepinephrine, propanolol, phentolamine and isoproterenol, did not inhibit antibody binding of labeled prostaglandin. Steroids which were free of significant cross-reactivity included digoxin, hydrocortisone, progesterone, estriol, testosterone and aldosterone. On the other hand, cholesterol (10^{-5}M) inhibited the anti-PGA_1-(^3H) PGA_1 reaction 18 percent. Unlabeled PGA_1, itself, totally inhibits binding of labeled PGA_1 at concentrations lower than 10^{-7}M, or greater than two orders of magnitude less. Cholesterol does not significantly inhibit at 10^{-7}M; however, since such high concentrations of cholesterol are present in physiologic fluids, the problem of cholesterol must be further investigated and dealt with. Table 3–III lists the degree of inhibition exhibited by several long chain fatty acids. The saturated fatty acids (caprilic, lauric, palmitic and arachidic) all behave similarly; chain length is a definite

TABLE 3–III

DEGREE OF INHIBITION OF BINDING OF (^3H) PGE, BY LONG CHAIN FATTY ACIDS

Fatty Acid (10^{-5} M)	*Degree of Inhibition (%)*
Caprilic	27.3
Lauric	30.6
Palmitic	33.8
Arachidic	33.8
Linoleic	71.0
Arachidonic	87.1

but minor factor. It is likely that some of this inhibition is not specific and may represent interference with the antigenantibody reaction by a detergent-like effect. In the biosynthesis of prostaglandins linoleic acid is converted to arachidonic acid and thence to prostaglandin. The close structural similarity of these unsaturated fatty acids with that of prostaglandins is reflected by a high degree of cross-reactivity.

TECHNIQUES OF PROSTAGLANDIN RADIOIMMUNOASSAY

For the sake of completeness, the exact details of the current technique for prostaglandin radioimmunoassay are described. In 12 x 75 mm polypropylene test tubes, 0.05 ml of antibody (1:1000 to 1:5000 dilution of serum), 0.05 ml of carrier rabbit gamma globulin (10 mg/ml), (^3H) PGE_1 (0.05 picomoles in 0.1 ml) and either known concentrations of prostaglandins or the unknown sample (in 0.1 ml) are incubated for seventy-two hours at 4° C. An equal volume of cold saturated ammonium sulfate is then added with stirring, incubated for thirty minutes

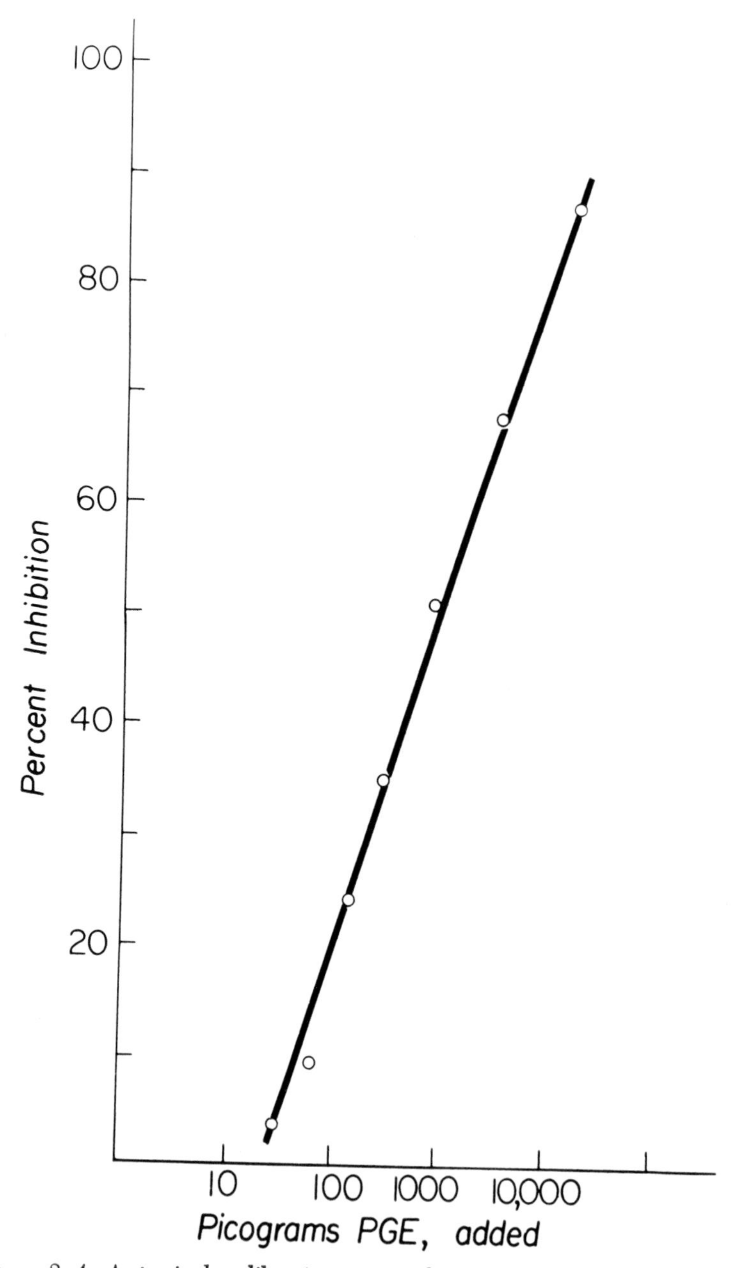

Figure 3–4. A typical calibration curve for measuring PGE_1 by radioimmunoassay. The plot is steep, linear on a semilog scale, and can be extrapolated to total inhibition.

and then centrifuged for thirty minutes at 3000 rpm. After the supernatant is decanted and discarded, the precipitate is washed with 50 percent saturated ammonium sulfate. The washed precipitate is dissolved in 0.5 ml of Nuclear Chicago Solubilizer, incubated one hour at 37° C, and then transferred quantitatively into toluene scintillation fluid in which it is counted.

A typical calibration curve for the radioimmunoassay utilizing an anti-PGE_1 antibody is illustrated in Figure 3–4. The maximal sensitivity of this immunoassay for PGA_1 and PGE_1 is now 10 picograms and, in addition, subpicomolar amounts of PGA_2 and PGE_2 can be detected. Techniques for the immunoassay of PGF compounds are currently being developed.

APPRAISAL

The application of radioimmunoassay to the prostaglandins threatens to make obsolete all previous biologic and chemical techniques for their measurement. The fantastic improvement in sensitivity should permit evaluation of the role of prostaglandins in physiologic situations. There are, however, three problems to be dealt with. First, the antibodies used to date are specific for prostaglandins but do cross-react to varying degrees with unsaturated fatty acids. Second, antibodies to one prostaglandin binds all prostaglandins, but to varying degrees. In order to measure each prostaglandin independently either the prostaglandins would have to be chemically separated before assay or perhaps measured simultaneously by a battery of antibodies and the values for each individual determined by solving complex simultaneous equations. Finally, application of this technique to measurement in serum or plasma or tissues may be difficult. The firm binding to albumin creates a serious obstacle and will probably require development of a useful means for extracting prostaglandins from serum. Extraction procedures described [41-44] are not reproducible enough, have only moderate recovery, and do not alleviate the problems offered by a nonspecific inhibitor. Thus, although a significant advance has been made in the development of the technique, there is still a great deal of work to be done.

REFERENCES

1. Kurzrok, R., and Lieb, C.C.: Biochemical studies of human semen. The action of semen on the human uterus. *Proc Soc Exp Biol Med, 28:*268, 1930.
2. Goldblatt, M.W.: A depressor substance in seminal fluid. *Chem Ind (Lond) 52:*1056, 1933.
3. von Euler, U.S.: Uber die spezifische blutdrucksenkende substanz des menschlichen prostata-und samenblasenkretes. *Klin Wochenschr, 14:*1182, 1935.
4. Bergstrom, S., and Sjovall, J.: The isolation of prostaglandin F from sheep prostate glands. *Acta Chem Scand, 14:*1693, 1960.
5. Bergstrom, S., Ryhage, R., Samuelsson, B., and Sjovall, J.: The structure of prostaglandin E, F_1 and F_2. *Acta Chem Scand, 16:*501, 1962.
6. von Euler, U.S., and Eliasson, R.: *Prostaglandins.* New York, Academic Press, 1968.
7. Bergstrom, S., Carlson, L.A., and Weeks, J.R.: The prostaglandins: a family of biologically active lipids. *Pharmacol Rev, 20:*1, 1968.
8. Anggard, E.: The isolation and determination of prostaglandins in lungs of sheep, guinea pig, monkey, and man. *Biochem Pharmacol, 14:*1507, 1965.
9. Anggard, E., and Bergstrom, S.: Biological effects of an unsaturated trihydroxy acid (PGF_{2a}) from normal swine lung. *Acta Physiol Scand, 58:*1, 1963.
10. Eliasson, R.: Studies on prostaglandin; occurrence, formation and biological actions. *Acta Physiol Scand, 46*(158):1, 1959.
11. Coceani, F., and Wolfe, L.S.: Prostaglandins in brain and the release of prostaglandin-like compounds from the cat cerebellar cortex. *Can J Physiol Pharmacol, 43:*445, 1965.
12. Horton, E.W., and Main, I.H.M.: A comparison of the biological activities of four prostaglandins. *Br J Pharmacol, 21:*182, 1963.
13. Sullivan, T.J.: Response of the mammalian uterus to prostaglandins under differing hormonal conditions. *Br J Pharmacol, 26:*678, 1966.
14. Ambache, N.: Irin and a hydroxy-acid from brain. *Biochem Pharmacol, 12:*421, 1963.
15. Lee, J.B., Covino, B.G., Takman, B.H., and Smith, E.R.: Renomedullary vasodepressor substance, medullin. *Circ Res, 17:*57, 1965.
16. Hickler, R.B., Lauler, D.P., Saravis, C.A., Vagnucci, A.I., Steiner, G., and Thorn, G.W.: Vasodepressor lipid from the renal medulla. *Can Med Assos J, 90:*280, 1964.
17. Strong, C.G., Boucher, R., Nowaczynski, W., and Genest, J.: Renal vasodepressor lipid. *Mayo Clin Proc, 41:*433, 1966.

18. Kannegiesser, H., and Lee, J.B.: Difference in haemodynamic response to prostaglandins A and E. *Nature, 229:*498, 1971.
19. Shaw, J.E., and Ramwell, P.W.: Separation, identification and estimation of prostaglandins. *Methods Biochem Anal, 17:*325, 1969.
20. Thompson, C.J., Los, M., and Horton, E.W.: The separation, identification and estimation of prostaglandins in nanogram quantities by combined gas chromatography-mass spectrometry. *Life Sci, 9:*983, 1970.
21. Anggard, E., and Samuelsson, B.: Purification and properties of a 15-hydroxyprostaglandin dehydrogenase from swine lung. *Arkh Kemi, 25:*293, 1966.
22. Anggard, E., Matschinsky, F., and Samuelsson, B.: Enzymatic assay of the prostaglandins. *Br J Pharmacol, 34:*190P, 1968.
23. Jaffe, B.M., Smith, J.W., Newton, W.T., and Parker, C.W.: Radioimmunoassay for prostaglandins. *Science, 171:*494, 1971.
24. Marrazzi, M.A., and Matschirsky, F.M.: Properties of 15-hydroxy prostaglandin dehydrogenase. *Pharmacology, 1:*373, 1972.
25. Samuelsson, B.: Prostaglandins and related factors. 27. Synthesis of tritium-labeled prostaglandin E_1 and studies on its distribution and excretion in the rat. *J Biol Chem, 239:*4091, 1964.
26. Andersen, N.H.: Preparative thin-layer and column chromatography of prostaglandins. *J Lipid Res, 10:*316, 1969.
27. Newton, W.T., McGuigan, J.E., and Jaffe, B.M.: Radioimmunoassay of peptides lacking tyrosine. *J Lab Clin Med, 75:*886, 1970.
28. Jaffe, B.M., Newton, W.T., and McGuigan, J.E.: Antibodies to gastrin tetrapeptide: a new technique for radioimmunoassay. *Fed Proc, 28:*501, 1969.
29. Goodfriend, T.L., and Ball, D.L.: Radioimmunoassay of bradykinin: chemical modification to enable use of radioactive iodine. *J Lab Clin Med, 73:*501, 1969.
30. Midgley, A.R., Jr., Niswender, G.D., and Ram, J.S.: Hapten-radioimmunoassay: a general procedure for the estimation of steroidal and other haptenic substances. *Steroids, 13:*731, 1969.
31. Levine, L., and Van Vunakis, H.: Antigenic activity of prostaglandins. *Biochem Biophys Res Commun, 41:*1171, 1970.
32. Greenwood, F.C., Hunter, W.M., and Glover, J.S.: The preparation of I-131 labelled human growth hormone of high specific radioactivity. *Biochem J, 89:*114, 1963.
33. Scatchard, F.: The attractions of proteins for small molecules and ions. *Ann NY Acad Sci, 51:*660, 1949.
34. Sheehan, J.C., and Hlavka, J.J.: The use of water-soluble and basic carbodiimides in peptide synthesis. *J Org Chem, 21:*439, 1956.
35. Jaffe, B.M., Newton, W.T., and McGuigan, J.E.: The effect of carriers on the production of antibodies to the gastrin tetrapeptide. *Immunochemistry, 7:*715, 1970.

36. Greenstein, J.P., and Winitz, M.: *Chemistry of the Amino Acids.* New York, Wiley, vol. 2, p. 978, 1961.
37. Oliver, G.C., Jr., Parker, B.M., Brasfield, D.L., and Parker, C.W.: The measurement of digitoxin in human serum by radioimmunoassay. *J Clin Invest, 47:*1035, 1968.
38. Parker, C.W., Gott, S.M., and Johnson, M.C.: The antibody response to a 2, 4-dinitrophenyl peptide. *Biochemistry, 5:*2314, 1966.
39. Parker, C.W., Gott, S.M., and Johnson, M.C.: Fluorescent probes for the study of antibody-hapten reaction. II. Variation in the antibody combining site during the immune response. *Biochemistry, 6:*3417, 1967.
40. Spragg, J., Schroder, E., Stewart, J.M., Austen, K.F., and Haber, E.: Structural requirements for binding to antibody of sequence variants of bradykinin. *Biochemistry, 6:*3933, 1967.
41. Sutherland, E. W., Robison, G.A., and Butcher, R.W.: Some aspects of the biological role of Adenosine 3′, 5′-monophosphate (Cyclic AMP). *Circulation, 27:*279, 1968.
42. Samuelsson, B.: Isolation and identification of prostaglandins from human seminal plasma. *J Biol Chem, 238:*3229, 1963.
43. Holmes, S.W., Horton, E.W., and Stewart, M.J.: Observations on the extraction of prostaglandins from blood. *Life Sci, 7:*349, 1968.
44. Hickler, R.B.: The identification and measurement of prostaglandin in human plasma. In Ramwell, P.W., and Shaw, J.E. (Eds.): *Prostaglandin Symposium of the Worchester Foundation of Experimental Biology.* New York, Wiley Interscience, p. 279, 1968.
45. Williams, E.D., Karim, S.M.M., and Sandler, M.: Prostaglandin secretion by medullary carcinoma of the thyroid. *Lancet, 1:*22, 1968.

CHAPTER 4

DIGITALIS: NEW LOOK AT AN OLD DRUG, ASSAY AND CLINICAL EXPERIENCE

GAETANO BAZZANO AND GAIL SANSONE BAZZANO*

INTRODUCTION

DIGITALIS IS INDEED an old drug: it was only 1775 when Withering learned from an old granddame in Shropshire that foxglove was good for dropsy.[1] He immediately set about trying it in heart diseases, recommending its use where he could and introducing it to the Edinburgh Pharmacopoeia in 1783. Shortly thereafter, his *Account of the Foxglove* (1785), a pharmacological classic, was a protest against use, misuse and abuses of digitalis, which were already creeping in.

Even though almost two hundred years have passed, under-use and/or abuse is still a contemporary problem with these extremely useful and powerful drugs. Clinicians have always been and still are continuously faced with a critical decision: whether to give a particular patient more of the drug, less of the drug or to stop the digitalis administration altogether. This decision may, at times, be a very difficult one, as demonstrated by the high incidence of cases of digitalis intoxication. Toxic manifestations of digitalis administration persist as a prevalent complication of cardiac therapy.[2-4]

The incidence of digitalis intoxication in hospitalized patients receiving digitalis has been reported to range from 8 to 20 per-

* The authors wish to thank Messrs. M. Gray and D. Eggerding for their technical assistance, Misses Carole Williams and Shannon Smith for the preparation of the manuscript, and the medical and paramedical personnel of the hospitals for their cooperation.

cent,[4-7] while the mortality as a direct result of cardiac toxicity has ranged from 3 to 21 percent.[3,5,6] The major difficulty of the above problem arises from the fact that both excessive and insufficient amounts of digitalis can produce identical signs and symptoms in the patient.

The need for information which would enable the physician to make a more rational decision with respect to the use of digitalis is reflected in the significant number of publications appearing in recent years on the determination of digitalis blood levels.

The usefulness of serum digitalis levels in clinical practice was first suggested by studies with tritiated digoxin, which indicated that therapeutic activity appeared to parallel the serum levels of the drug.[8] Earlier studies using radiocarbon digoxin [9-11] had not supported the usefulness of such knowledge, probably because of the small number of patients investigated.

Several methods have hence been developed, during the last five years, for the determination of the plasma concentration of cardiac glycosides (i.e. digitoxin and digoxin) for clinical application. Prerequisites for any such method, if it is to have widespread clinical application, are:

1. the length of time necessary to run an assay must be short enough to make the results available in time for a change in therapeutic direction;
2. it must make use of readily available equipment; and
3. it has to have a degree of sensitivity and specificity sufficient to distinguish between small but critical variations in plasma digitalis levels.

These requirements preclude the usefulness and applicability of a myriad of chemical, biological, chromatographic and double isotopic techniques which have been used in the past.

At the present time, in our opinion, there are three assays that provide for real clinical application.

1. ENZYMATIC ASSAY

Described first by Burnett et al.[12] and more recently applied by Bentley et al.,[13] this assay involves (a) extraction of digitoxin from plasma by an organic solvent, (b) evaporation of an aliquot of extract to dryness in a test tube and (c) assay of transport adenosine triphosphate (ATP-ase) *in vitro* where enzymatic activity is determined by phosphate release. Enzyme inhibition by dried digitoxin in the tube is dose-dependent, and this is the basis for the assay. ATP-ase enzyme is prepared from homogenized hog brain through a series of differential centrifugations and iodide extractions. A high speed refrigerated centrifuge is required for this preparation. The authors have found plasma maintenance levels of digitoxin comparable to those reported by Lukas and Peterson[14] and Oliver et al.[15] Although there was some overlap of groups, levels appeared to correlate well with toxicity degrees. This method, according to the authors, seems to have met the requirements of a simple, practical and accurate assay of cardiac glycoside levels.[13] The test is currently not applicable to digoxin.

2. RADIOIMMUNOASSAY

This now widely used assay, independently described by Smith et al.[16,17] and Oliver et al.,[15,18] appears to be at this time the most sensitive, and can be completed within a working day. In this assay, tritiated digoxin and digoxin-specific antibodies are added to a plasma sample; the digoxin present in the plasma will compete with the known quantity of tritiated digoxin for antibody binding. The percentage of tritiated digoxin bound to antibody is then a function of that digoxin present in the plasma (unknown) sample. The assay is also used for the determination of digitoxin. With this radioimmunoassay, digoxin and digitoxin plasma levels measured in patients believed to be digitalis-intoxicated are much higher than levels obtained in patients judged as adequately digitalized; however, a wide range is noted in "adequately digitalized subjects." The cause of this variability is not clear as yet.

3. ^{86}Rb RED CELL UPTAKE

The last and most widely used technique for measuring both digoxin and digitoxin levels is the ^{86}Rb uptake assay of Lowenstein and Corrill.[19] In this procedure, the digitalis glycosides inhibit the uptake of ^{86}Rb by red blood cells. The assay consists of (a) the extraction of digitoxin or digoxin with methylene chloride, (b) evaporation to dryness, (c) incubation at 37° with fresh packed red cells, (d) the addition of ^{86}Rb and (e) measurement of the red cell uptake, using a γ-counter at a level of 10 to 12 percent efficiency. The ^{86}Rb uptake by the red cell is quantitatively inhibited by the glycosides present in the methylene chloride extract. Some of the drawbacks inherent in this technique are: the necessity of a standard dose response curve for each assay performed, a lack of accuracy for digoxin plasma concentrations below 0.5 ng/ml of plasma and accessibility of a γ-counter. Finally, there is a wide range of values (as in other assays) accepted as in the therapeutic range and considerable overlap between toxic and nontoxic patients. In spite of the drawbacks, we have found the ^{86}Rb uptake assay to be extremely useful in clinical practice.

Over the past two years, we have determined the plasma digitoxin levels in 150 patients who were judged to be adequately digitalized and not intoxicated according to clinical signs and symptoms. Our results showed a mean digitoxin level of 26.45 ng/ml $\pm$ 7.9 SD. The mean digoxin level in 50 patients, who were also judged to be adequately digitalized and not intoxicated, was 2.62 ng/ml $\pm$ 1.2 SD. Our results using this method are quite similar to those reported for digoxin by Grahame-Smith and Everest.[20]

The determination of plasma digitalis levels has been used in our laboratory to study the effect of colestipol,* a nonabsorbable bile acid binding polymer, on digitalis blood levels in patients presenting with symptoms and signs of digitalis intoxication.

It has long been known that digitalis glycosides undergo significant biliary excretion and intestional reabsorption.[21,22] This concept has been recently reconfirmed by Caldwell, who demonstrated in rats that after the intraduodenal administration of ^{3}H-digoxin or ^{3}H-digitoxin, 45 to 50 percent of the labeled glycosides are excreted in the bile within the first twenty-four hours.[23]

* Colestipol (U26,597A), an insoluble copolymer of tetraethylenepentamine and epicholorohydrin, was supplied by the Upjohn Co., Kalamazoo, Mich.

In humans and under normal, nontoxic conditions, digoxin appears to have a lesser enterohepatic circulation than digitoxin.[24]

Due to the structural similarity of digitalis with bile acids, these resins should be capable of binding digitalis preparations in the enterohepatic circulation, shorten their physiologic half-life and cause a decrease in digitalis blood levels. Studies on the binding capacity of colestipol and cholestyramine † for digitalis glycosides *in vitro* are also reported.

MATERIALS AND METHODS

Six patients were admitted to the Intensive Cardiovascular Care Unit of the St. Louis University Hospital or the St. Louis City Hospital Unit II Medical Service with the clinical diagnosis of digitalis intoxication. In accordance with the Helsinki Declaration, an informed consent was obtained from each patient.

Four patients with digitoxin toxicity and one patient with digoxin toxicity received colestipol, 10 gm stat. and 5 gm q. 6 to 8 hours PO. Two other patients served as controls for each one of the groups. Cardiac glycosides were discontinued in all patients.

Digoxin and digitoxin serial plasma levels were determined using the ^{86}Rb red cell uptake assay previously discussed.[19,20] ^{86}RbC1 was obtained from Amersham-Searle Corporation, Des Plaines, Illinois. Radioactivity counts were obtained using a Packard automatic γ-counter at a level of 10 to 12 percent efficiency. Methylene chloride and RbC1 were purchased from Fisher Scientific Company, St. Louis, Missouri.

Electrocardiographic monitoring, electrolytes and other laboratory procedures were done as routinely carried out in the above hospitals, using standard clinical laboratory techniques.

The glycoside binding capacity of both cholestyramine and colestipol was studied *in vitro* without and with the addition of duodenal juice obtained and pooled from three normal subjects after secretin stimulation. It was thought that the addition of

† Cholestyramine, a quanternary ammonium styrenedivinylbenzene copolymer, was supplied by the Upjohn Company, Kalamazoo, Michigan, 49001.

duodenal juice might approximate conditions of the proximal small intestine, where binding *in vivo* should presumably take place.

For the *in vitro* determination of the glycoside binding capacity, a resin column was prepared with 1 gm of cholestyramine or colestipol. After washing, the radioactive glycoside containing 50 ug/100,000 cpm was added to the column and eluted with deionized water in 10 ml fractions. It was found that most of the radioactivity was contained in the eluate fractions 1, 2 and 3, rapidly decreasing in 4 and no radioactivity in fraction 5. The eluates 1 through 5 were pooled, rechecked for radioactivity, extracted with methylene chloride and rechecked for radioactivity. The biological activity, with the ^{86}Rb assay, was also determined. The resin was divided and half was washed with 5 ml of deionized water, extracted with methylene chloride and the methylene chloride extract checked for radioactivity and the biological activity determined. The other half of the resin was treated with 5 ml of 0.1 N HC1 and extracted with methylene chloride. This fraction was then rechecked for radioactivity and biological activity.

RESULTS AND DISCUSSION

The results of the *in vitro* studies comparing the binding capacity of colestipol and cholestyramine are presented in Table 4–I. Colestipol was capable of binding 4.8 ± .20 ng of digoxin and 7.4 ± .27 ng of digitoxin per mg of resin. Cholestyramine, on the other hand, was able to bind 6.4 ± .40 ng/mg of digoxin and

TABLE 4–I
IN VITRO BINDING CAPACITY

	Digoxin *ng/mg ± SEM* *	*Digitoxin* *ng/mg ± SEM* *
Colestipol	4.8 ± 0.20	7.4 ± 0.27
Colestipol +D.J. †	4.7 ± 0.73	5.1 ± 0.42
Cholestyramine	6.4 ± 0.40	14.4 ± 1.28
Cholestyramine +D.J.	2.8 ± 0.42	4.6 ± 1.08

* Mean ± standard error of mean.
† Duodenal juice.

14.4 ± 1.28 ng/mg of digitoxin. These results are compatible with the *in vitro* binding capacity of the resins for cholic acid: colestipol binds approximately 1 mEq cholic acid per gram,[25] while cholestyramine, determined under similar conditions, binds 2.2 mEq/gm of resin.[26]

From these results it would appear that cholestyramine is the resin to be used in digitoxin or digoxin toxicity because of its significantly greater binding capacity. However, when the resins were premixed and incubated at 37° for one hour with duodenal juice, colestipol showed a binding capacity of 4.7 ± .73 ng/mg (Mean ± SEM) for digoxin and 5.1 ± .42 ng/mg for digitoxin. Cholestyramine plus duodenal juice bound 2.8 ± .42 ng/mg of digoxin and 4.6 ± 1.08 ng/mg of digitoxin. Therefore, the presence of duodenal juice decreases the binding capacity of cholestyramine, while it does not seem to have a significant effect on the binding capacity of cholestipol.

The decrease in the binding capacity of cholestyramine was 56 percent for digoxin and 69 percent for digitoxin, while no change in the colestipol-binding capacity was seen for digoxin and the decrease for digitoxin was 31 percent. The differences, which are significant ($p < .05$) in the binding of digoxin and digitoxin by the two resins in the presence of duodenal juice, do not seem to be related to changes in pH nor to the bile acid content of the duodenal juice. The pH of our incubation mixtures for these studies was compatible with the pH present in the small intestine, ranging between 6.5 and 7.5. Extreme pH changes, 1 to 3 and 10 to 12 respectively, influence the binding capacity; this, of course, should be expected since these are ion exchange resins. The duodenal juice was obtained from several patients, pooled and the aliquots added to the incubation mixture taken from an homogenous pool. Temperature was also kept constant and all incubations were carried out at 37°.

A possible explanation for this behavior could be the displacement of bile acid by the cardiac glycosides from the binding sites of colestipol and not of cholestyramine (more strongly bound), or a steric hindrance due to the presence of bile acids complexed in greater quantity with the cholestyramine.

Because of the above results, colestipol and not cholestyra-

TABLE 4–II
DEGREES OF DIGITALIS TOXICITY

Grade I.	Slight toxicity:	Anorexia, ventricular ectopic beats, bradycardia.
Grade II.	Moderate toxicity:	Nausea and vomiting, headache, malaise, ventricular premature beats.
Grade III.	Severe toxicity:	Diarrhea, blurring of vision, confusion, disorientation, ventricular tachycardia, SA or AV block.
Grade IV.	Extreme toxicity:	Abdominal pain, high-degree conduction blocks, and atrial or ventricular fibrillation.

mine has been used for the treatment of patients presenting with digitalis toxicity.

For the purpose of this study, digitalis toxicity has been graded as shown in Table 4–II, and patients retrospectively classified accordingly.

Figure 4–1 presents a semilog diagram of plasma digitoxin levels in one control patient and four patients treated with colestipol (at various time intervals). The degree of intoxication for each patient, classified as per Table 4–II, is indicated. The serum digitoxin as well as the electrolyte levels are listed in Table 4–III. As we can see from Figure 4–1 and Table 4–III, starting levels of plasma digitoxin were, with one exception (L.T.), greater than 45 ng/ml. Although the patient (L.T.) presented with a history of nausea and emesis and the admission ECG showed ventricular premature contractions in the form of a bigeminal rhythm with atrial fibrillation, an initial plasma digitoxin level of only 35 ng/ml was found. A subsequent plasma level lower than 20 ng/ml suggested discontinuation of colestipol and administration of 0.2 mg of digitoxin PO. This case, once more, underlines the reported considerable overlap in serum digitalis levels in toxic and nontoxic groups of patients.[27]

As can be seen from Table 4–III, the serum potassium levels were not appreciably decreased in this nor any other patient. Prevalence of hypokalemia has been reported in patients with digitalis toxicity,[28,29] although no other authors found significant difference in serum potassium between toxic and nontoxic digitalized patients.[17,30] No constant relationship between hypokalemia and the toxicity grading was evident from our data. The same lack of relationship holds true in respect to the glycoside levels and the state of toxicity, a finding supported by other investigators, who have reported considerable overlap in

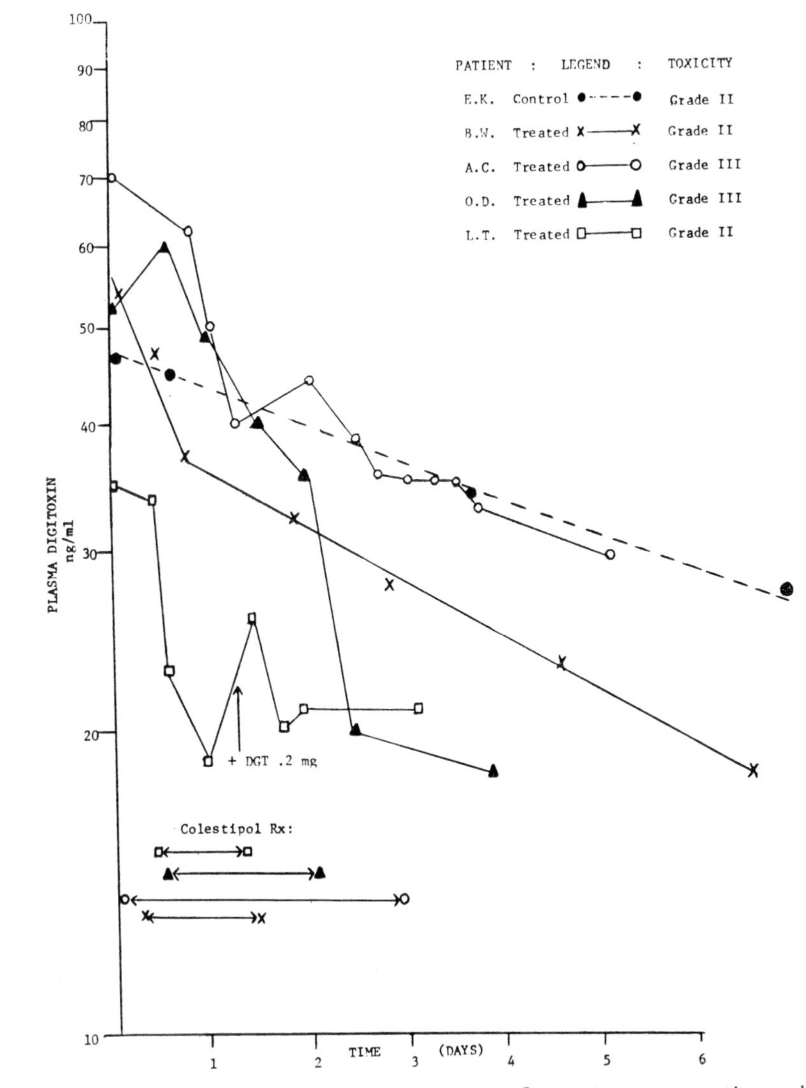

Figure 4–1. Serial plasma levels in five study patients presenting with clinical signs of digitoxin intoxication.

TABLE 4–III

GLYCOSIDE AND SERUM ELECTROLYTE LEVELS IN A CONTROL AND COLESTIPOL TREATED SUBJECTS PRESENTING WITH DIGITOXIN INTOXICATION

Patient	*Time hours*	*DGT ng/ml*		*Na^+ mEq/liter*	*K^+ mEq/liter*	*Cl^- mEq/liter*	*$^-CO_2$ mEq/liter*
E.K.	0	46		—	—	—	—
	12	45		—	—	—	—
	84	35		146	4.1	108	20
	180	28		138	4.5	92	25
B.W.	0	52		—	—	—	—
	8	47	Rx	133	5.0	90	20
	17	37	│	131	5.1	90	29
	41	32	│	128	5.8	91	31
	65	28	↓	128	4.8	90	29
	161	18					
L.T.	0	35		141	4.2	100	—
	10	34	Rx	130	3.4	98	24
	22	18	│	137	5.0	100	24
	33	26	↓	138	4.6	94	23
	45	21		133	4.4	94	25
A.C.	0	70	Rx	145	4.1	98	27
	19	62	│	144	4.1	101	27
	30	41	│	144	4.2	102	26
	60	38	│	144	4.4	106	22
	65	36	│	—	—	—	—
	72	35	│	—	—	—	—
	84	36	↓	144	4.5	107	27
	90	36		135	4.0	96	28
O.D.	0	52		148	4.6	101	—
	14	60	Rx	149	4.9	101	—
	24	40	│	145	5.3	102	—
	36	49	│	157	6.5	111	—
	46	36	↓	143	5.5	98	—
	60	19		141	5.4	100	—

Rx = Colestipol 10 gm stat followed by 5 gm q̄ 5 to 6 hours. Other supportive measures equal in all five subjects.

serum digitalis levels in toxic and nontoxic groups of patients.[27]

As can be seen from Figure 4–1, the administration of colestipol resulted in a rapid and marked decrease in the serum levels of digitoxin. Clinical signs of digitalis toxicity disappeared within the first 24 to 30 hours, and the electrocardiographic changes reverted towards normal. By comparison, the digitoxin levels in the untreated control patient fell gradually and clinical and electrocardiographic evidence of toxicity persisted until the third day after admission, at which time the plasma digitoxin level was approximately 37 ng/ml.

Figure 4–2 presents serial plasma levels for two patients, A.B. and H.W., admitted with digoxin intoxication. Both patients were classified as Grade III toxicity. Plasma digoxin levels were 17 and 14 ng/ml respectively. A.B., considered as a control, was treated by supportive measures. Colestipol administration in H.W. was started only the second day after admission.

The decrease in the digoxin plasma level of H.W. appears to be dramatic when compared to that of the control patient (A.B.) treated only with supportive measures. From the observation of the slopes of the two curves, it could be inferred that when colestipol administration was discontinued (approximately 16 hours from the loading dose), the fractional turnover coefficient (rate of disappearance) of the glycoside from the plasma compartment was equivalent (day 3 to day 6) in both patients.

In a recent review of the problems and management of digi-

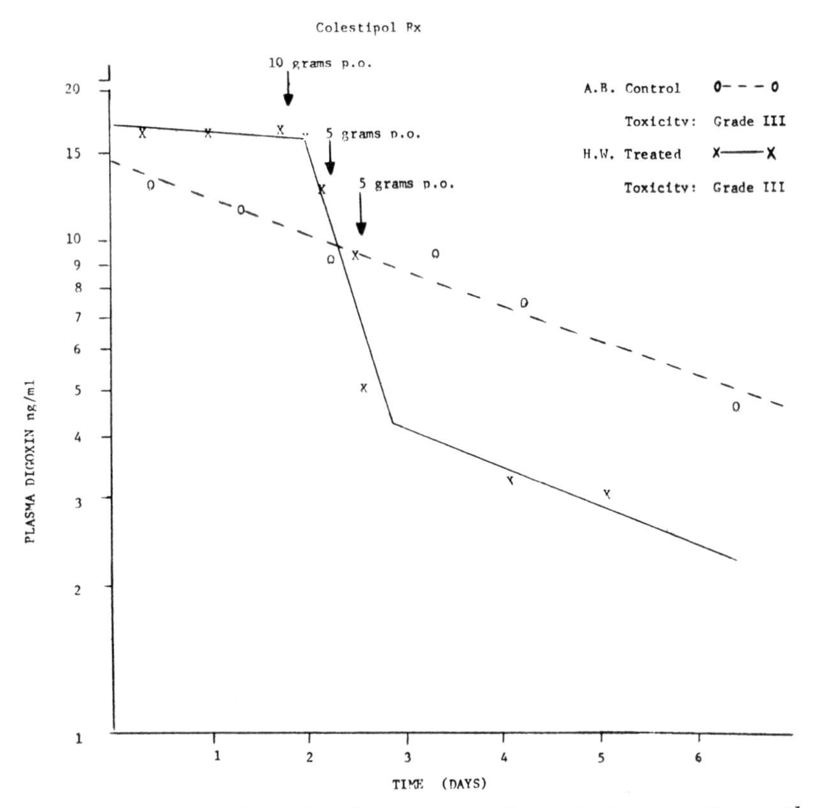

Figure 4–2. Serial plasma levels in two study patients presenting with clinical signs of digoxin intoxication.

talis intoxication, Butler[31] suggested that an immunologic approach to the reversal of toxicity was the present hope; however, we do not share this author's hopes. To us, it appears obvious that a treatment utilizing antibodies, even if they were available in sufficient amounts, would involve hazards to the patient, including the potential of serum sickness. Furthermore, the question of whether or not, in the myocardium, digitalis is mainly intracellular or in a membrane-bound form is not known; therefore, it is not proven that antibodies would be effective even if they became available.

Since resins are readily obtainable and administrable[23,32] and since their action is rapid and can be obtained without systemic absorption, they appear to be the most convenient, safe and effective method for the treatment of digitalis toxicity presently known.

SUMMARY

The most common current techniques for plasma digitalis determination in clinical practice have been reviewed. The binding capacity of colestipol, a new anion exchange resin, and cholestyramine for digoxin and digitoxin has been studied. *In vitro* experiments have shown that the presence of duodenal juice may decrease the binding capacity of cholestyramine while it does not seem to have a significant effect on the binding capacity of colestipol. Colestipol was administered to patients presenting signs and symptoms of digoxin or digitoxin intoxication. Two patients—one with digoxin and one with digitoxin intoxications—used as controls, were treated with supportive measures only.

Clinical signs and plasma glycoside levels returned to normal in a much shorter time in the patients treated with colestipol. It appears therefore that colestipol may be of significant usefulness in the treatment of digitalis intoxication.

REFERENCES

1. Garrison, F.H.: *History of Medicine*, 4th ed. Philadelphia, W. B. Saunders, p. 357, 1929.

2. Dreifus, L.S., McKnight, E.H., Katz, M., and Likoff, W.: Digitalis intolerance. *Geriatrics, 18*:494, 1963.
3. Herrmann, G.R.: Digitoxicity in the aged: recognition, frequency, and management. *Geriatrics, 21*:109, 1966.
4. Shapiro, S., Slone, D., Lewis, G.P., and Hershel, J.: The epidemiology of digoxin: a study in three Boston hospitals. *J Chronic Dis, 22*:361, 1969.
5. Rodensky, P.L., and Wasserman, F.: Observations on digitalis intoxication. *Arch Intern Med, 108*:171, 1961.
6. Schott, A.: Observations on digitalis intoxication—a plea. *Postgrad Med J, 40*:628, 1964.
7. Gotsman, M.S., and Schrire, V.: Toxicity—a frequent complication of digitalis therapy. *South African Med J, 40*:590, 1966.
8. Doherty, J.E., Perkins, W.H., and Mitchell, G.K.: Tritiated digoxin studies in human subjects. *Arch Intern Med, 108*:531, 1961.
9. Okita, G.T., Kelsey, F.E., Talso, P.J., Smith, L.B., and Geiling, E.M.K.: Studies on the renal excretion of radioactive digitoxin in human subjects with cardiac failure. *Circulation, 7*:161, 1953.
10. Okita, G.T., Kelsey, F.E., Walaszek, E.J., and Geiling, E.M.K.: Biosynthesis and isolation of Carbon-14 labeled digitoxin. *J Pharmacol Exp Ther, 110*:244, 1954.
11. Okita, G.T.: Studies with radioactive digitalis. *J Am Geriatr Soc, 5*:163, 1957.
12. Burnett, G.H., and Conklin, R.L.: The enzymatic assay of plasma digitoxin levels. *J Lab Clin Med, 71*:1040, 1968.
13. Bentley, J.D., Burnett, G.H., Conklin, R.L., and Wasserburger, R.H.: Clinical application of serum digitoxin levels. *Circulation, 41*:67, 1970.
14. Lukas, D.S., and Peterson, R.E.: Double isotope dilution derivative assay of digitoxin in plasma, urine and stool of patients maintained on the drug. *J Clin Invest, 45*:782, 1966.
15. Oliver, G.C., Jr., Parker, B.M., Brasfield, D.L., and Parker, C.W.: The measurement of digitoxin in human serum by radioimmunoassay. *J Clin Invest, 47*:1035, 1968.
16. Smith, T.W., Butler, V.P., Jr., and Haber, E.: Determination of therapeutic and toxic serum digoxin concentrations by radioimmunoassay. *N Engl J Med, 281*:1212, 1969.
17. Smith, T.W.: Radioimmunoassay for serum digitoxin concentration: methodology and clinical experience. *J Pharmacol Exp Ther, 175*:352, 1970.
18. Oliver, G.C., Parker, B.M., and Parker, C.W.: Radioimmunoassay for digoxin: technique and clinical application. *Am J Med,* in press.
19. Lowenstein, J.M., and Corrill, E.M.: An improved method for measuring plasma and tissue concentrations of digitalis glycosides. *J Lab Clin Med, 67*:1048, 1966.
20. Grahame-Smith, D.G., and Everest, M.S.: Measurement of digoxin

in plasma and its use in diagnosis of digoxin intoxication. *Br Med J, 1*:286, 1969.

21. Lauterbach, F.: Enterale Resorption, biliare Ausscheidung und entero-hepatischer Kreislaul von Herzglykosiden bei der Ratte. Naunyn Schmiedeberg. *Arch Exp Pathol, 247*:391, 1964.
22. Okita, G.T.: Species difference in duration of action of cardiac glycosides. *Fed Proc, 26*:1125, 1967.
23. Caldwell, J.H., and Greenberger, N.J.: Cholestyramine enhances digitalis excretion and protests against lethal intoxication. *J Clin Invest, 49*:16A, 1970.
24. Doherty, J.E., Flanigan, W.J., Murphy, M.L., Bulloch, R.T., Dalrymple, G.L., Beard, O.W., and Perkins, W.H.: Tritiated digoxin: XIV. Enterohepatic circulation, absorption, and excretion studies in human volunteers. *Circulation, 42*:867, 1970.
25. Parkinson, T.M., Gundersen, K., and Nelson, N.A.: Effects of colestipol (U-26,597A), a new bile acid sequestrant, on serum lipids in experimental animals and man. *Atherosclerosis, 11*:531, 1970.
26. Tennent, D.M., Siegel, H., Zanetti, M.E., Kuron, G.W., Ott, W.H., and Wolf, F.J.: Plasma cholesterol lowering action of bile acid binding polymers in experimental animals. *J Lipid Res, 1*:469, 1960.
27. Beller, G.A., Smith, T.W., Abelmann, W.H., Haber, E., and Hood, W.B.: Digitalis intoxication: a prospective clinical study with serum level correlations. *N Engl J Med, 284*:989, 1971.
28. Soffer, A.: The changing clinical picture of digitalis intoxication. *Arch Intern Med, 107*:681, 1961.
29. Shrager, M.W.: Digitalis intoxication: a review and report of forty cases, with emphasis on etiology. *Arch Intern Med, 100*:881, 1957.
30. Smith, T.W., and Haber, E.: Digoxin intoxication: the relationship of clinical presentation to serum digoxin concentration. *J Clin Invest, 49*:2377, 1970.
31. Butler, V.P., Jr.: Digoxin: immunologic approaches to measurement and reversal of toxicity. *N Engl J Med, 283*:1150, 1970.
32. Bazzano, G., Gray, M., and Sansone-Bazzano, G.: Treatment of digitalis intoxication with a new steroid-binding resin. *Clin Res, 18*:592, 1970.

CHAPTER 5

SATURATION ANALYSIS OF THYROID HORMONE

Helmut Haibach

INTRODUCTION

There are two thyroid hormones: thyroxine (T_4) and triiodothyronine (T_3). Thyroxine circulates with a serum concentration approximately thirty times greater than triiodothyronine.[1,2] This explains why tests measuring circulating thyroid hormone were first developed for thyroxine and later for triiodothyronine. The earlier tests for thyroxine were chemical determinations,[1] difficult and time consuming procedures. They are now being replaced by faster methods based on saturation analysis for which radionuclides are used. Since radio chemicals have become readily available to many laboratories, these tests have gained much in popularity. Saturation analysis is applied to the determination of thyroxine and triiodothyronine. There are several different methods. I shall describe and discuss all methods briefly. The method of Murphy and Pattee will be examined in more detail since it lends itself to the discussion of important aspects of saturation analysis.

PRINCIPLE OF SATURATION ANALYSIS

The assay system consists of a binding agent (B), the ligand (S) and the radioactively labeled ligand (S*). The ligand is the compound to be measured. The labeled and unlabeled ligand compete with each other to form complexes with the binding agent: BS and BS*. Both reactions follow the mass law (Fig. 5–1) and are allowed to progress to equilibrium or near equilibrium. The equilibrium is governed by the reaction constants k_1, k_2, k_3 and k_4. The parameter measured in the system is the distribution of the radioactive label between BS* and S* at

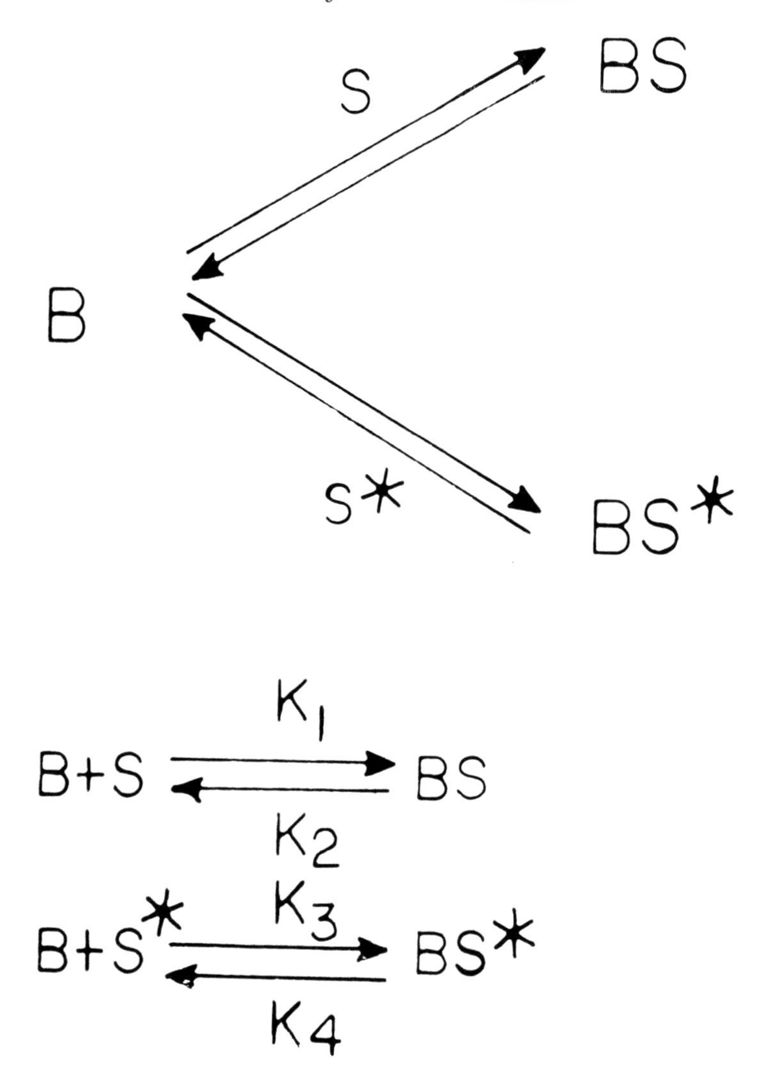

Figure 5–1. Principle of saturation analysis. B = Binding agent; S = Ligand = Compound to be measured; S* = Labeled ligand; BS = Ligand binding agent complex; BS* = Ligand* binding agent complex.

equilibrium or near equilibrium. Complete separation of BS* and S* is therefore necessary. The more S is present in the system, the higher is the ratio S*/BS*, the less S the lower the ratio. A standard curve allows conversion of S*/BS* into quantities of S. Suitable concentrations of the binding agent are selected that give approximately equal distribution of the label between S* and BS* for the expected average of the quantities of S to be assayed.

Different names have been used for assays with this principle. The term "Saturation Analysis" was introduced by Barakat and Ekins [3] in 1961. It emphasizes that the S*/BS* measures the saturation of the binding agent. It is the most inclusive term, but gives no information with regard to the binding agent. Murphy [4] proposed "Competitive Protein Binding" analysis. This term is more specific in as much as it identifies the binding agent as protein. The assay protein for thyroxine determination is human TBG (thyroxine binding globulin). The term is most applicable to the assays of nonantigenic hormones like thyroxine, testosterone and other compounds as folic acid and vitamin B_{12}; many of these methods were developed in the same laboratory.[5] Earlier, Murphy [5] had suggested "Radiostereo Assay" [3] to point out that the ligand has a certain steric configuration for which the binding site of TBG is highly specific. Robbins and Rall [6] suggested "Displacement Analysis." It refers to the technique of Barakat and Ekins [3,7] which is described below. In this method the displacement of thyroxine from one assay protein to another, from TBG to albumin, is measured; Korenman [8] added "Radioligand Binding Assay" to the list. The term conveys neither the idea of saturation nor of specificity. "Radioimmuno Assay" [9] is also based on the principle described above. There are distinct differences between competitive protein binding and radioimmunoassay. Both assays have been discussed in the literature [9] under the common heading "Saturation Analysis." This does not appear practical any more because of the rapid expansion of the field of immunoassay and its special problems. Radioimmunoassay has been discussed in earlier chapters in depth. It will be dealt with here briefly, in as much as it has been applied to thyroid hormone measurement.

For the purpose of this review, competitive protein binding assay differs from radioimmunoassay in that the former makes use of physiological serum or tissue proteins as assay proteins. They have a high binding specificity and association constants. Radioimmunoassay utilizes specific antibodies with greater specificity and much greater association constants.

THYROXINE DETERMINATION

The Murphy-Pattee Method

The first report of this method—a competitive protein binding technique—appeared in 1964.[10] It became a useful laboratory tool after its major modification [11] in 1965. There are three major steps. In the first, the sample separation, the serum proteins are precipitated with ethanol and thyroxine is extracted. The second step involves equilibration of the dried extracted thyroxine with the tracer and the assay protein. The third step is separation of the bound and unbound ligand for determination of the distribution of the radioactivity. In detail, 1 ml of serum is mixed with 2 ml 95% ethanol. After thorough mixing and centrifugation 0.3 ml of the supernatant are dried with an air jet at 45° C. The residue is dissolved in 1 ml barbital buffer, 500 ml of which contain 5 to 25 μCi ^{131}I-thyroxine, 15 ml pooled serum with the assay protein, phenol and propyleneglycol. Isotopic equilibration is achieved by shaking for eight minutes at 45° C. The mixture is cooled before the resin for the absorption of the unbound ligand is added. It settles quickly after addition of more cold barbital buffer. Two milliliters of the supernatant is radioassayed and compared to the total radioactivity of the sample before addition of the resin. A standard curve with four points is obtained by processing ethanol solutions with 5, 10, 20 and 30 ng of unlabeled T_4.

The Assay Protein

Pooled human serum is the source of the assay protein. It contains at least three thyroxine binding proteins, thyroxine-binding globulin (TBG), albumin (A) and thyroxine-binding prealbumin (TBPA) [12] (Fig. 5–2). Each has a different binding capacity and affinity for thyroxine. TBG has the most suitable properties: the highest affinity and greatest specificity. It is therefore most suited to serve as the assay protein. Thyroxine binding by TBPA is inhibited by barbital buffer and binding by albumin was reduced by dilution. The concentration of the assay protein in the system is important. It determines the slope and

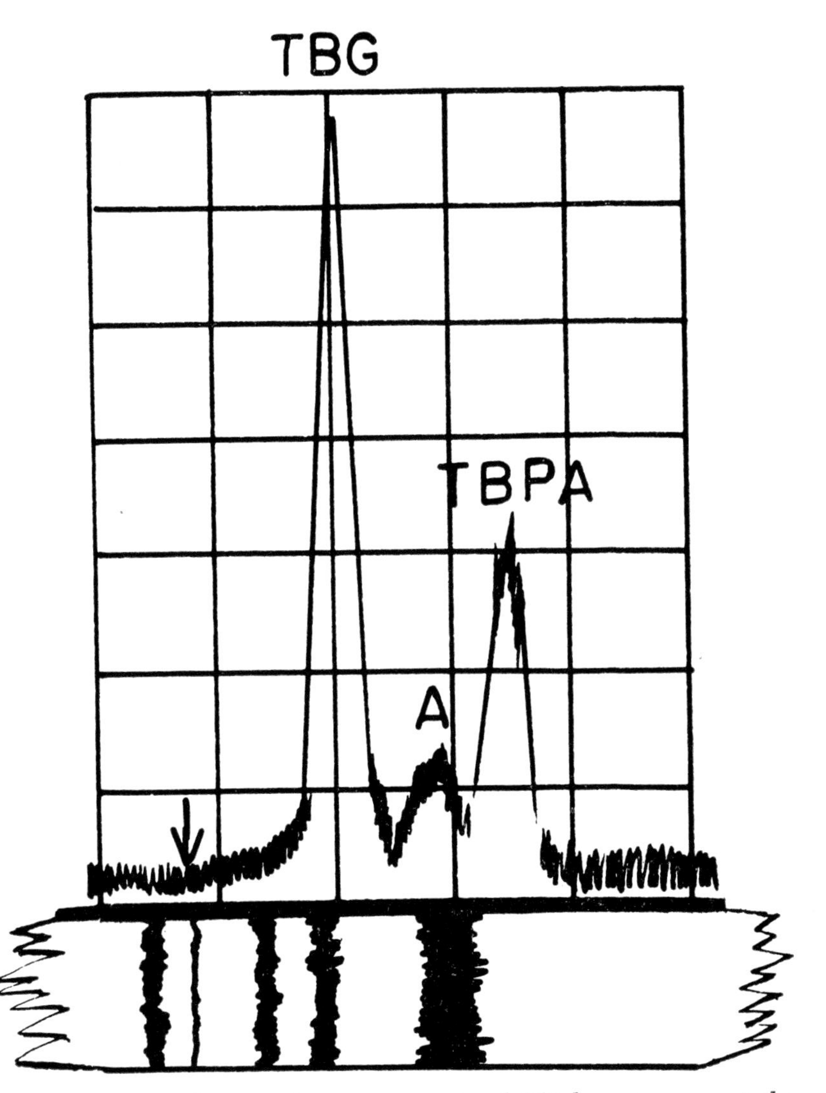

Figure 5–2. Distribution of tracer amounts of ^{131}I-thyroxine among the serum proteins as determined by paper electrophoresis. A strip stained for protein is shown in the lower panel. Arrow indicates point of application. TBG = Thyroxine-binding globulin; A = Albumin; TBPA = Thyroxine-binding prealbumin. (From J.H. Oppenheimer, *N Engl J Med, 278:*1153, 1968.)

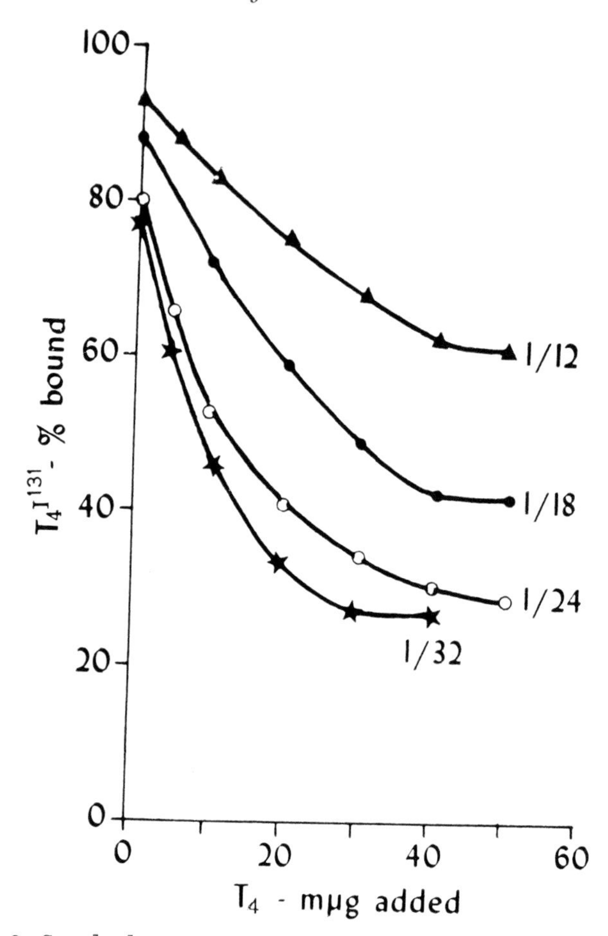

Figure 5–3. Standard curves using various serum dilutions. The percentage bound ^{131}I–T_4 is plotted versus mμg unlabeled T_4 added. (From B.E.P. Murphy, and C.J. Pattee, *J Clin Endocrinol Metab, 24:*187, 1964.)

degree of linearity of the standard curve and also the range of possible measurements (Fig. 5–3). The lowest of the curves shown is obtained with a serum dilution of 1/32. This concentration gives the steepest curve for the desired range of measurements, 0 to 20 ng. The range covers almost all serum thyroxine concentrations observed in normal and abnormal thyroid function. The temperature of the reagent mixture on the thyroxine binding by TBG has been carefully studied (Fig. 5–4). Lower temperatures have a moderately beneficial effect on the standard

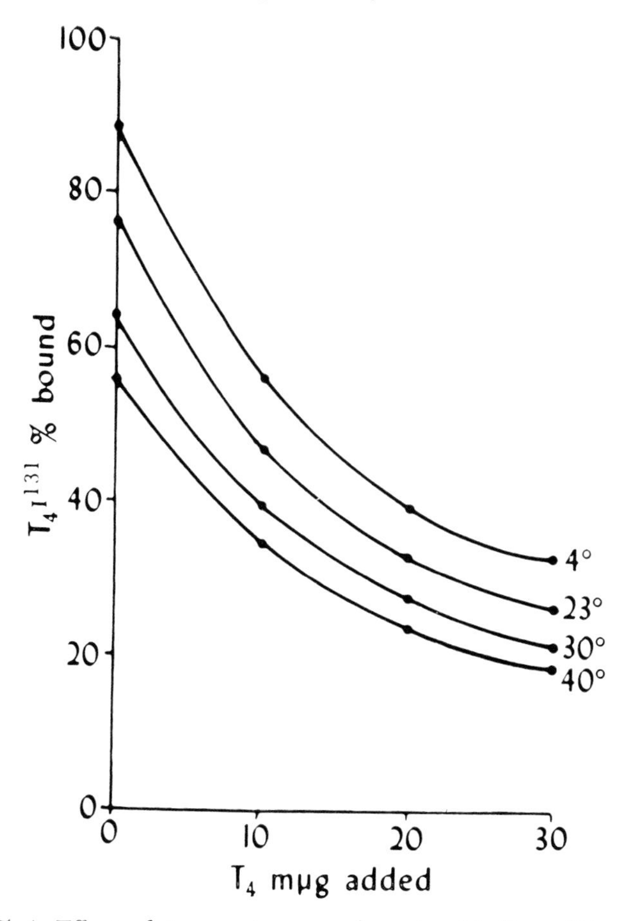

Figure 5–4. Effect of temperature on thyroxine binding. (From B.E.P. Murphy, and J.C. Pattee, *J Clin Endocrinol Metab, 24:*187, 1964.)

curve, increasing the linearity for the most frequently used range and increasing the slope. What is gained in sensitivity for the assay appears small in comparison with the additional effort that cooling requires.

Specificity of the Method

The binding of thyroxine by TBG is highly specific. The steric configuration of the thyroxine molecule with regard to serum protein-binding has been studied in detail.[13] There are no other known physiological compounds in the human blood that bind to TBG at the thyroxine site except for triiodothyronine which binds much less firmly (10:1) and which is present in much lower concentrations (30:1) in serum [2] and thus does not interfere with the thyroxine determination. Diphenylhydantoin when circulating in therapeutic levels lowers the concentration of circulating thyroxine, but it does not compete significantly with thyroxine in the assay system though it is extracted by ethanol together with thyroxine. Murphy [5] has examined diphenylhydantoin-containing sera and found comparable data using both her assay and the PBI. Dextrothyroxine, a nonphysiologic stereoisomer of the hormone levothyroxine, is occasionally administered to patients; it increases the assay results but exerts little or no hormonal effect when circulating, thus giving misleading high values.

Precision

Precision of the method depends largely on the slope and range of the standard curve. Murphy [5] sets as a rule of thumb that the percentage-bound ligand must drop at least one half, preferably two thirds over the range to be measured. In her laboratory λ is 0.2 or less with a mid-range coefficient of variation usually of 10 percent. Duplicate or triplicate determinations reduce this error to 6 or 7 percent, a quite acceptable level.

Accuracy

The accuracy of the procedure hinges much on the first step. Thyroxine extraction from the serum sample is incomplete but consistent at 77 ± 4.8 percent. This would indicate that the method underestimates the circulating thyroxine concentration. The given normal range 4 to 11 μg/100 ml serum compared with that of the PBI 4 to 8 μg/100 ml appears high. There are several factors to be considered: 65.4 percent of thyroxine is iodine; some thyroxine is lost in the extraction procedure of either method. Fitzgerald et al.[14] have analyzed sera from a large group of patients with suspected thyroid disease employing both tests and found that the results are in very good agreement when appropriate corrections for loss in thyroxine extraction are made for both assays.

Frequently Encountered Problems

The most serious problem of the method is related to the deproteinization of the serum sample. Serum proteins transferred into the assay system interfere with TBG binding. Mailed serum samples, likely to be exposed to repeated rapid temperature changes, are believed to undergo some protein denaturation which presumably accounts for some falsely high results. The method has been used but has not been sufficiently tested for animal serum,[15] such as rat serum. Deterioration of the unlabeled thyroxine standard will also give falsely high results, an often perplexing but easily resolved problem. Labeled thyroxine preparations [15] are not infrequently contaminated with tracer iodide; iodide exceeding 10 percent will give erroneous high results. Chromatographic analysis of the commercially available preparations is indicated from time to time. Misleading low results can be obtained when part of the extracted thyroxine is destroyed during evaporation of the ethanol. A forceful air jet probably facilitates oxidation; nitrogen jets do not degrade thyroxine under these conditions. Gentle air jets are satisfactory.

Clinical Evaluation

Murphy et al.[16] have supplied ample data establishing the usefulness of their method. The normal range was determined in three major series (Fig. 5–5). Impressive is the good correlation with the PBI (Fig. 5–6). (The difference between the mean of the protein-bound iodine of 5.8 and the mean thyroxine iodine of 4.1 μg/100 ml as shown in the figure is discussed above.) Sera from a wide range of patients has been examined, with normal thyroid function, hyper- and hypothyroidism, increased and decreased TBG levels, and with other conditions. The method also provides useful information when the PBI is invalid because of contaminating iodine (radiological contrast media, surgical preps, or iodine-containing medications) and iodide, an increasingly more frequent situation in clinical medicine. This is a major advantage. PBI levels are falsely low shortly after treatment with mercurials. Mercury in the serum interferes with the chemical reaction. In this situation, the Murphy-Pattee method also gives

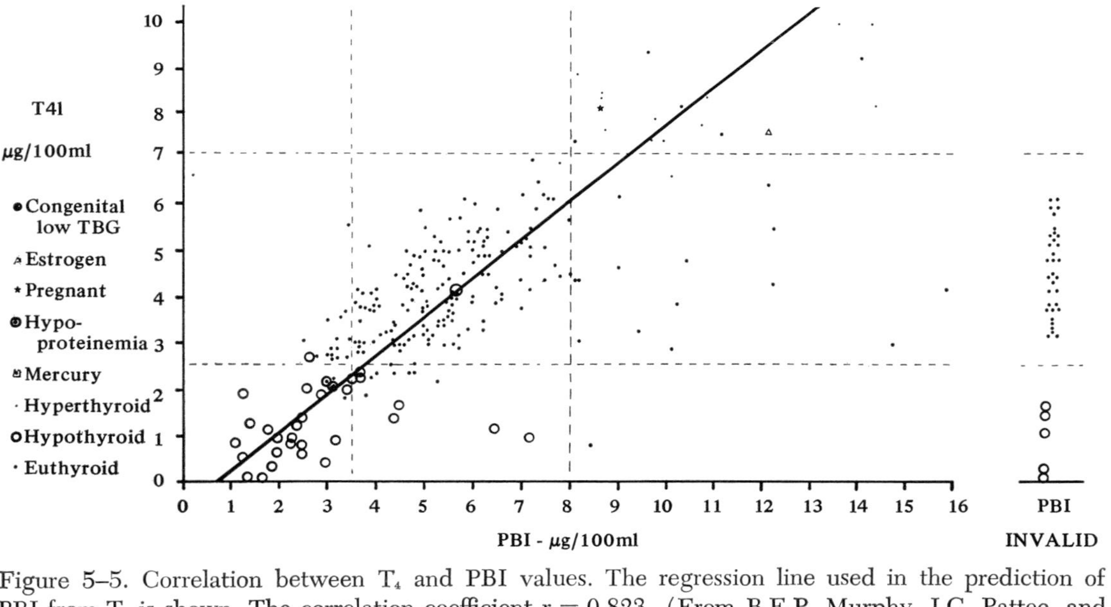

Figure 5–5. Correlation between T_4 and PBI values. The regression line used in the prediction of PBI from T_4 is shown. The correlation coefficient $r = 0.823$. (From B.E.P. Murphy, J.C. Pattee, and A. Gold, *J Clin Endocrinol Metab, 26*:247, 1966.)

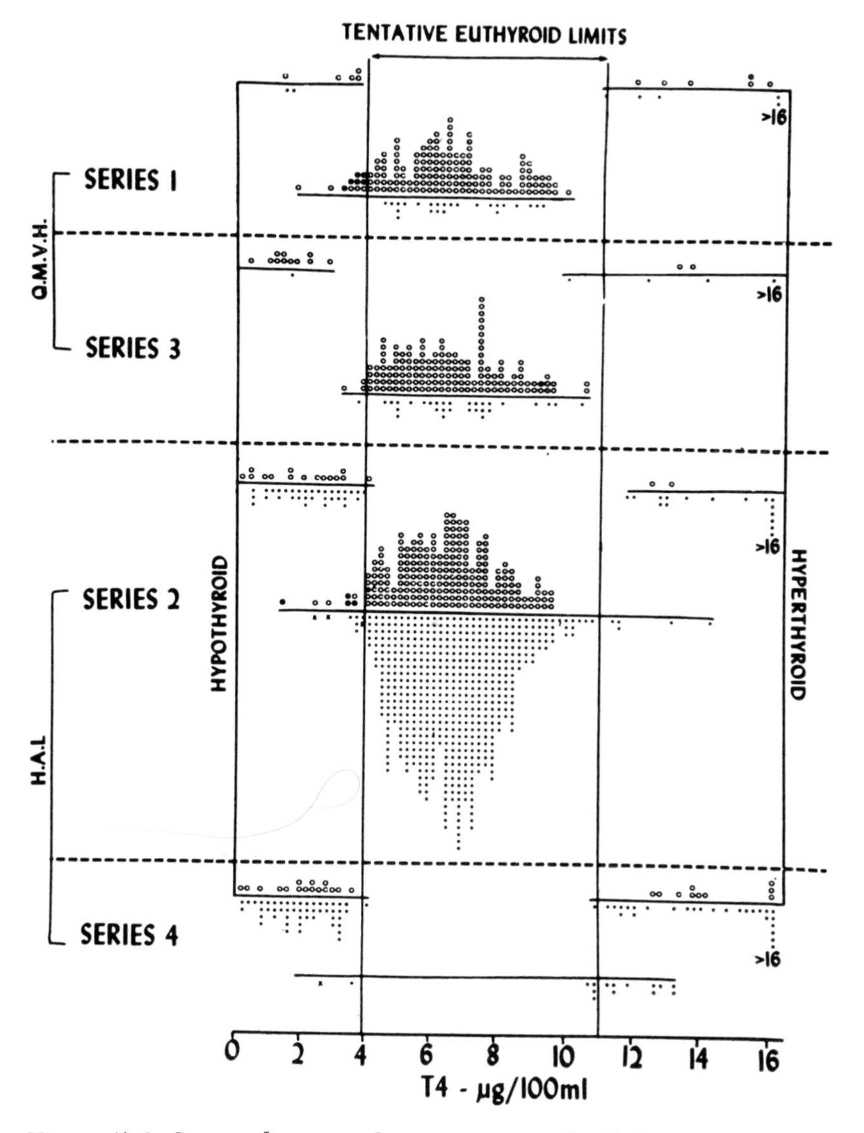

Figure 5–6. Serum thyroxine data on 1146 individuals, as related to adopted limits of "normal." Hypothyroid values appear on left, euthyroid in middle, and hyperthyroid on right. Males are indicated by open circles above lines, females by dots below lines; crosses indicate euthyroid subjects with low values and hypoproteinemia. (From B.E.P. Murphy, J.C. Pattee, and A. Gold, *J Clin Endocrinol Metab*, *26*:247, 1966.)

reliable results. Other frequently encountered situations were examined as shown in Figure 5–7. A small but statistically significant difference between the sexes is demonstrated for serum thyroxine (Fig. 5–8). As expected from the slightly higher TBG levels, females have higher thyroxine levels. The observation is of little clinical importance, but does add to the credibility of the assay.

The method has worked well in other laboratories and proved equally useful.[17,18] Cassidy et al.[17] point out that the competitive protein-binding assay gave reliable results for sera with very

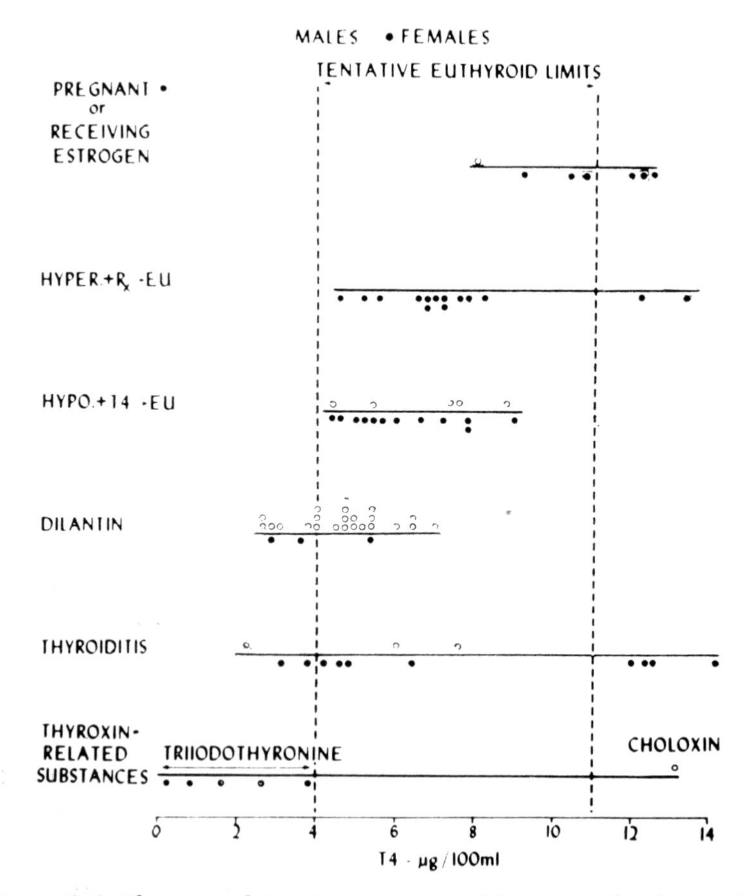

Figure 5–7. Thyroxine determination in special groups of subjects. The layout is similar to that of Figure 5–6. (From B.E.P. Murphy, J.C. Pattee, and A. Gold, *J Clin Endocrinol Metab, 26*:247, 1966.)

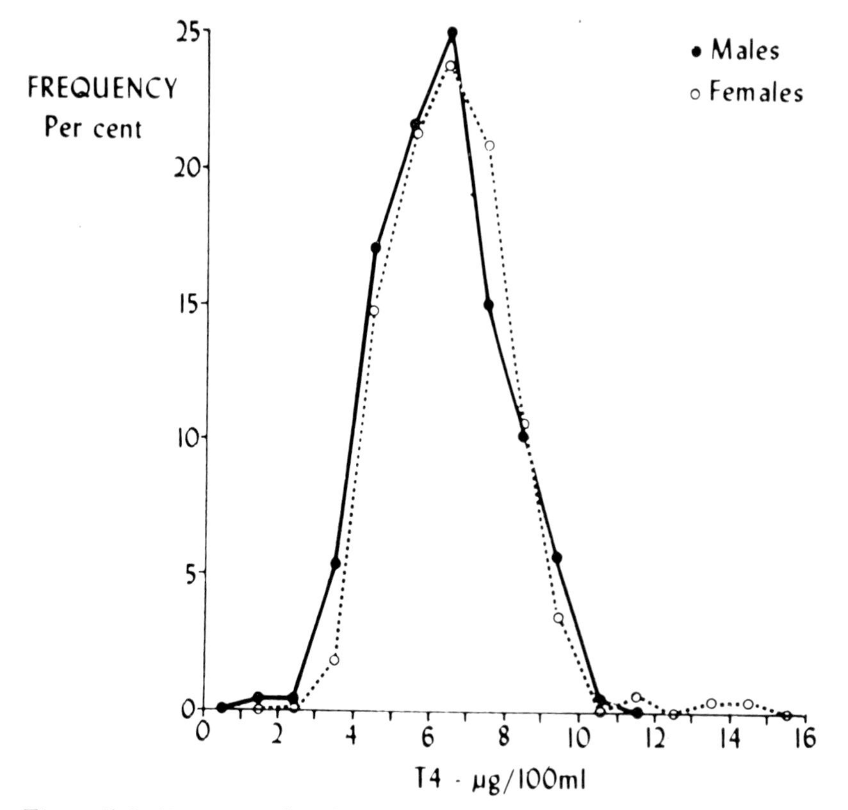

Figure 5–8. Frequency distributions for males and females. (From B.E.P. Murphy, J.C. Pattee, and A. Gold, *J Clin Endocrinol Metab, 26*:247, 1966.)

high organic iodine contamination when the "T_4-by-column" method which is designed for iodine removal gives falsely high values because of incomplete separation of the contaminant. Miller [19] has compared the Murphy-Pattee method with the twenty-four-hour thyroid readioiodide uptake tests in 140 patients examined for eu-, hyper and hypothyroidism. He reports a high correlation between the two tests.

Serum thyroxine determination is now carried out in many laboratories using "kits" from different commercial sources employing competitive protein binding. The methods are quick and economical and proved valuable for screening purposes.[20,21] A recent study [20] compared two of these methods. The authors claim greater accuracy and a smaller range of normal values for one of them but the difference is not very convincing. It appears that studies with much larger numbers of serum samples examining various kit procedures, the PBI and the method as described by Murphy and Pattee have to be carried out before more valid comparisons can be made.

Other Methods

The first method, applying the principle of saturation analysis to thyroxine determination, was reported by Barakat and Ekins in 1961.[3] They equilibrated serum samples with tracer doses of $^{131}I-T_4$ and extracted with ethanol, monitoring the recovery of the label. The dry residue of the extract was mixed with normal human serum. The distribution of the label between TBG and albumin was correlated with thyroxine standards. This method proved too complicated for routine use but deserves credit for the pioneer work. Nakajima et al.[22] described a method using ethanol precipitation and extraction. The dried supernatant was mixed with normal plasma in tris buffer allowing TBG- and TBPA-binding. They measured displacement from TBG to an anion exchange resin. The results were compatible with the clinical diagnosis in eighty-two patients of theirs.

Recently and since this symposium was held, a report on a radioimmunoassay for thyroxine by Chopra, Solomon and Ho [23] appeared. The assay employs T_4-binding antisera which are produced in rabbits immunized with human thyroglobulin in Freud's adjuvant. Serum T_4 measured by competitive protein binding and this new technique was highly correlated in a large series of samples. The data from this laboratory indicate that radioimmunoassay for thyroxine is feasible and is comparable or preferable to competitive protein binding in accuracy, specificity, sensitivity, efficiency and technician time.

DETERMINATION OF THYROXINE-BINDING CAPACITY OF SERUM

Thyroxine and triiodothyronine in normal human serum are bound to specific binding proteins except for a very small free or dialyzable fraction, 0.046 percent for thyroxine [12] and 0.46 percent for triiodothyronine.[24] The free hormone concentration in serum corresponds well with the level of thyroid function and regulates thyrotropin secretion.[1] Free thyroid hormone levels are therefore more informative than total hormone concentrations. However, all available techniques measuring free dialyzable hormone [12,25,26] are not yet suitable for routine use. Thyroxine-binding capacity of the serum proteins reflects the percentage unbound hormone. It can be easily measured with the "T_3-resin uptake," a technique that has been used widely and successfully.[9,27]

In 1957, Hamolsky et al.[28] found that red cells bind a greater proportion of labeled thyroxine in oxalated blood from thyrotoxic subjects than red cells in normal blood. They developed a red cell uptake test [28,29] on the basis of this observation. Whole blood is equilibrated with $^{131}I\text{-}T_3$. The distribution of the label between red cells and plasma is determined after centrifugation and repeated washing of the cells with buffer. The binding capacity of the cells depends on their number, the blood pH and other factors.[9] Correction for the hematocrit complicates the procedure and does not reduce the variability sufficiently. The difficulty has been overcome by Sterling et al.[27] who developed an anion exchange resin uptake test. It was found to be simple and reliable.[9,30] One might ask why the test is conducted with T_3 rather than with T_4, since T_4 is the main circulating thyroid hormone. T_3 is bound to TBG, but much less firmly than T_4. The resin uptake of labeled T_3 is around 35 percent. T_4 which is bound much more firmly would give lower ratios. Such ratios are much more subject to error secondary to incomplete separation of serum and resin. Small amounts of TBG-bound tracer will give falsely high ratios. Kits are now available for this test from several commercial sources.

Results from this test can be combined with those of the PBI or Murphy-Pattee-T_4 determination for calculation of the free thyroxine index, an indicator of the absolute concentration of circulating free hormone. This value helps interpret test results in conditions with high or low TBG concentrations.[1] However, a more important reason for the popularity of the T_3-resin uptake test is that it gave useful information for iodine contamination serum, while it was the only such test readily available before introduction of the Murphy-Pattee method and the "T_4-by-column" test.

DETERMINATION OF SERUM TRIIODOTHYRONINE

Since its discovery by Gross and Pitt-Rivers [31] in 1952, 3, 5, 3′-triiodothyronine has received increasing attention. Though its precise place in thyroid hormone physiology and pathophysiology is not known, calculations based on its blood concentration of total and free hormone, its faster turnover in the periphery [32] and its higher metabolic activity [1] suggest that it plays a role in tissue metabolism approximately equal to that of thyroxine. The newly described disorder of T_3 toxicosis,[33] its importance in maintaining euthyroidism after treatment of hyperthyroidism,[33a] and the data indicating peripheral [34] and intrathyroidal [35] conversion of thyroxine to triiodothyronine stimulated the interest in methods for triiodothyronine determination.

Reports on radioimmunoassay for triiodothyronine have recently appeared.[36-38,40,41] Brown et al.[39] provoked antibody production by coupling T_3 to succinylated poly-l-lysine. Gharib et al.[36]

TABLE 5–I
METHODS FOR MEASUREMENT OF CIRCULATING TRIIODOTHYRONINE

Assay Developed by	*Also Used by*	*Normal Range*	*Technique*
Naumann, Naumann and Werner [24]	—	330 ± 70 ng/100 ml M ± SD	CPB
Sterling, Bellabarba, Newman and Brenner [2]	—	220 ± 27 ng/100 ml M ± SD	PC CPB
Sterling et al.	Wahner and Gorman [46]	243 ± 40 ng/100 ml M ± SD	PC CPB
Sterling et al.	Hotelling and Sherwood [47]	210 ± 20 ng/100 ml M ± SE	PC CPB
Mitsuma, Colluci, Shenkman and Hollander [41]	—	134 ± 6.2 ng/100 ml M ± SE	RIA
Lieblich and Utiger [45]	—	145 ± 25 ng/100 ml M ± SD	RIA
Larson, P. R.[44]	—	110 ± 25 ng/100 ml M ± SD	RIA
Chopra, Solomon and Beall [40]	—	100 − 170 ng/100 ml	RIA
Chopra and Lam [48]		109 ± 5.2 ng/100 ml	RIA
Bioscience Laboratories	—	60 − 190 ng/100 ml	RIA

PC = paper chromatographic separation of T_4 and T_3
CPB = competitive protein binding
RIA = radioimmunoassay
( ) = numbers in brackets are references

failed to obtain an appropriate response taking the same approach, but succeeded in developing a satisfactory assay using antibodies to a conjugate of T_3 and human, bovine or rabbit serum albumin. The group reports a sensitivity one thousand times greater than that of the other methods. They tested thirty derivatives of tyrosines and thyronines proving adequate specificity. Thyroxine which is present in normal serum in a concentration approximately thirty times that of triiodothyronine has a very low potency in the assay (0.05% to 0.01% of that of T_3) and therefore does not interfere with T_3 determination in spite of its higher concentration. Physiological studies initially conducted to prove the validity of the assay have been encouraging,[36] however, this method has become a useful tool fit for routine clinical [44] use. The method of Sterling et al.[2] and the double isotope derivative assay of Hagen et al.[41] do not lend themselves to routine use. Nauman et al.[24] modified the Murphy-Pattee method for T_3 measurements. However, their T_3 levels for euthyroid subjects (330 ± 70 ng/100 ml, mean $\pm$ SD) are high in comparison with those of the assays from Sterling's (220 ± 27) and Mayberry's (218 ± 55) laboratories. Even lower normal ranges were reported from several other laboratories (Table 5–I), all of them using radioimmunoassay methods. The ranges which most closely reflect the true situation have not been decided. Regardless of the range of normal, each assay provides the same clinically useful information.

SUMMARY

There are two widely used, simple and reliable methods with the principle of saturation analysis for thyroid hormone measurement: the Murphy-Pattee method, a competitive protein-binding assay, for the measurement of serum thyroxine; and the T_3-resin uptake test for assessment of the thyroxine-binding capacity. Both give useful information when other tests fail because of iodine contamination in serum, an increasingly frequent situation in clinical medicine. No sufficiently simple technique for triiodothyronine measurement is yet widely available. Radioimmunoassay procedures, recently described, are expected to fill this gap and perhaps replace the competitive protein-binding assay for thyroxine.

REFERENCES

1. Williams, R.H.: *Textbook of Endocrinology.* Philadelphia, Saunders, 1968.
2. Sterling, K., Ballabarba, D., Newman, E.S., and Brenner, M.A.: Determination of triiodothyronine concentration in human serum. *J Clin Invest, 48:*1150, 1969.
3. Barakat, R.M., and Ekins, R.P.: Assay of vitamin B_{12} in blood, a simple method. *Lancet, II,* 25, 1961.
4. Murphy, B.E.P.: Application of the property of protein-binding to the assay of minute quantities of hormones and other substances. *Nature (Lond), 201:*679, 1964.
5. Murphy, B.E.P.: Protein binding and the assay of nonantigenic hormones. *Rec Progr Hormone Res, 25:*563, 1969.
6. Gray, C.H., and Bacharach, A.L.: *Hormones in Blood.* New York, Academic Press, 1967.
7. Ekins, R.P.: The estimation of thyroxine in human plasma by an electrophoretic technique. *Clin Chim Acta, 5:*453, 1960.
8. Korenman, S.G.: Radio-ligand binding assay of specific estrogens using a soluble uterine macromolecule. *J Clin Endocrinol Metab, 28:*127, 1968.
9. Wagner, H.N., Jr.: *Principles of Nuclear Medicine.* Philadelphia, Saunders, 1968.
10. Murphy, B.E.P., and Pattee, C.J.: Determination of thyroxine utilizing the property of protein-binding. *J Clin Endocrinol Metab, 24:*187, 1964.
11. Murphy, B.P.: The determination of thyroxine by competitive protein-binding analysis employing an anion-exchange resin and radiothyroxine. *J Lab Clin Med, 66:*161, 1965.
12. Oppenheimer, J.H.: Role of plasma proteins in the binding distribution, and metabolism of the thyroid hormones. *N Engl J Med, 278:*1153, 1968.
13. Robbin, J., and Rall, J.E.: Proteins associated with the thyroid hormones. *Physiol Rev 40:*415, 1960.
14. Fitzgerald, L.T., Bruno, F.P., Glassman, A., and Williams, C.M.: Serum thyroxine determination by competitive protein binding analysis: the normal range. *J Nucl Med, 11:*669, 1970.
15. Haibach, H., and McKenzie, J.M.: Increased free thyroxine postoperatively in the rat. *Endocrinology, 81:*435, 1967.
16. Murphy, B.E., Pattee, C.J., and Gold, A.: Clinical evaluation of a new method for the determination of serum thyroxine. *J Clin Endocrinol Metab, 26:*247, 1966.
17. Cassidy, C.F., Benotti, J., and Peno, S.: Clinical evaluation of the determination of thyroxine iodine. *J Clin Endocrinol Metab, 28:*420, 1968.

18. Kennedy, J.A., and Abelson, D.M.: Determination of serum thyroxine using a resin sponge technique. *J Clin Pathol, 20:*89, 1967.
19. Miller, R.: Correlation of thyroid-function studies. *N Engl J Med, 283:*935, 1970.
20. Brookeman, V.A., and Williams, C.M.: Evaluation of the resin strip technique for determining serum T_3 binding capacity and serum thyroxine. *J Nucl Med, 12:*55, 1971.
21. Thorson, S.C., Tsujikawa, R., Brown, J.L., Morrison, R.T., and McIntosh, H.W.: Evaluation of a simplified method for determining serum thyroxine by competitive protein binding analysis. *Acta Endocrinol (Kbh), 64:*630, 1970.
22. Nakajima, H., Kuramochi, M., Horiguchi, T., and Kubo, S.: A new and simple method for the determination of thyroxine in serum. *J Clin Endocrinol Metab, 26:*99, 1966.
23. Chopra, I.J.: A radioimmuno assay for measurement of thyroxine in unextracted serum. *J Clin Endocrinol Metab, 34:*938, 1972.
24. Nauman, J.A., Nauman, A., and Werner, S.C.: Total and free triiodothyronine as a test of thyroid function. *J Clin Invest, 21:*456, 1961.
25. Ingbar, S.H., Braverman, L.E., Dawbar, N.A., and Lee, G.Y.: A new method for measuring the free thyroid hormone in human serum and an analysis of the factors that influence its concentration. *J Clin Invest, 44:*1679, 1965.
26. Sterling, K., and Brenner, M.A.: Free thyroxine in human serum: simplified measurement with the aid of magnesium precipitation. *J Clin Invest, 45:*155, 1966.
27. Sterling, K., and Tabachnick, M.: Resin uptake of ^{131}I-triiodothyronine as a test of thyroid function. *J Clin Endocrinol Metab, 21:*456, 1961.
28. Hamolsky, M.W., Stein, M., and Freedberg, A.S.: Thyroid hormone plasma protein complex in man. II. A new *in vitro* method for study of "uptake" of labeled hormonal components by human erythrocytes. *J Clin Endocrinol Metab, 17:*33, 1957.
29. Hamolsky, M.W., Golodetz, A., and Freedberg, A.S.: The plasma protein-thyroid hormone complex in man. III. Further studies on the use of the *in vitro* red blood cell uptake of ^{131}I-l-triiodothyronine as a diagnostic test of thyroid function. *J Clin Endocrinol Metab, 19:*103, 1959.
30. Sisson, J.C.: Principles of, and pitfalls in, thyroid function test. *J Nucl Med, 6:*853, 1965.
31. Gross, J., and Pitt-Rivers, R.: The identification of 3:5:3′-l-triiodothyronine in human plasma. *Lancet, 1:*439, 1952.
32. Sterling, K., Lashof, J.C., and Man, E.B.: Disappearance of serum ^{131}I-labeled l-thyroxine and l-triiodothyronine in euthyroid subjects. *J Clin Invest, 33:*103, 1954.
33. Sterling, K., Refetoff, S., and Selenkow, H.A.: T_3 thyrotoxicosis.

Thyrotoxicosis due to elevated serum triiodothyronine levels. *JAMA, 213:*571, 1970.

33a. Sterling, K., Brenner, M.A., Newman, E.S., Odell, W.D., and Bellabarba, D.: The significance of triiodothyronine maintenance of euthyroid states after treatment of hypothyroidism. *J Clin Endocrinol Metab. 33:*729, 1971.

34. Sterling, K., Brenner, M.A., and Newman, E.S.: Conversion of thyroxine to triiodothyronine in normal human subjects. *Science, 169:*1099, 1970.

35. Haibach, H.: Free iodothyronines in the rat thyroid gland. *Endocrinology, 88:*149, 1971.

36. Gharib, H., Mayberry, W.E., and Ryan, R.J.: Radioimmunoassay for triiodothyronine: a preliminary report. *J Clin Endocrinol Metab, 31:*709, 1970.

37. Gharib, H., Mayberry, W.E., Hocket, T., and Ryan, R.J.: Radioimmunoassay for triiodothyronine. *Program of the Annual Meeting of the Endocrine Society,* 1971, Abstract, 188.

38. Brown, B.L., Ekins, R.P., and Ellis, S.M.: A radioimmunoassay for serum triiodothyronine. *J Endocrinol, 46:*1, 1970.

39. Brown, B.L., Ekins, R.P., Ellis, S.M., and Reith, W.S.: Specific antibodies to triiodothyronine hormone. *Nature, 226:*359, 1970.

40. Chopan, T.J., Soloman, D.H., and Beall, G.N.: Radioimmunoassay for measurement of triiodothyronine in human serum. *J Clin Invest, 50:*2033, 1971.

41. Mitsuma, T., Colucci, J., Shinkman, L., and Hollander, C.S.: Rapid simultaneous radioimmunoassay for triiodothyronine and thyroxine in unextracted serum. *Biochem Biophys Res Comm, 46:*2107, 1972.

42. Hagen, G.A., Diuguid, L.I., Kliman, B., and Stanbury, J.B.: Double isotope derivative assay serum triiodothyronine. *Clin Res, 18:*602, 1970.

43. Larson, P.R.: Direct immunoassay of triiodothyronine in human serum. *J Clin Invest, 51:*1939, 1972.

44. Larson, P.R.: Triiodothyronine: Review of recent studies of its physiology and pathophysiology in man. *Metabolism, 21:*1073, 1972.

45. Lieblich, J., and Ntiger, R.D.: Triiodothyronine radioimmunoassay. *J Clin Invest, 51:*1939, 1972.

46. Wahner, H.W., and Gorman, C.A.: Interpretation of serum triiodothyronine levels measured by the Sterling Technique. *N Engl J Med, 284:*225, 1971.

47. Hotelling, R.D., and Sherwood, L.M.: The effects of pregnancy on circulating triiodothyronine. *J Clin Endocrinol Metab, 33:*783, 1971.

48. Chopra, I.J., and Lam, R.W.: Use of 8-anilino-1-napthaline-sulfonic acid (ANS) in radioimmunoassay of triiodothyronine in unextracted serum. *Clin Res, 20:*216, 1972.

CHAPTER 6

RADIOASSAY FOR VITAMIN B_{12}

Robert E. Henry and Helmut Haibach

INTRODUCTION

Rickes, Folkers and coworkers'[1] isolation of vitamin B_{12} (B_{12}) in 1948 prompted studies that unequivocally identified intestinal malabsorption of B_{12} as the cause of the manifestations of pernicious anemia.[2,3] Intrinsic factor deficiency was the first cause of defective B_{12} absorption recognized. Many other disorders interfering with B_{12} absorption, such as exocrine pancreatic deficiency, parasitic competition for B_{12} in the intestinal lumen and destruction of the ileal wall by various infections, neoplastic or other diseases were subsequently reported. Radioactively-labeled B_{12} has proven to be a useful tool for *in vivo* tracer studies of B_{12} absorption and distribution. Diagnostic procedures involving the administration of labeled B_{12} to patients, however, are time consuming and cumbersome. They require complete collections of stool or urine for a minimum of two to five days, and in some instances surface counting of the liver or total body counting. The most widely utilized procedure, the "Schilling Test,"[4] has the added disadvantage that the loading dose of nonradioactive B_{12} given during the test produces hematologic remission precluding a therapeutic trial with physiologic quanities of B_{12}. Procedures which measure the absorption of orally administered B_{12}, therefore, are better suited to defining the etiology of the malabsorption rather than determining the presence of a B_{12} deficiency state.

Because of the very low concentration of B_{12}, in the order of picograms/ml, in serum and tissues, chemical assays were not seriously attempted. Early estimations of B_{12} levels were performed with a microbiologic assay measuring B_{12}-dependent growth of microorganism. *Euglena gracilis* or *Lactobacillus leichmanii* served as test organisms.[5,6] Patients with B_{12} deficiency could be separated from healthy subjects and patients with other disorders on the basis of serum B_{12} concentration, thus establishing the microbiologic assay as a useful screening tool. However, this assay has several disadvantages. Most hospital laboratories find it inconvenient to maintain a stock culture of test organisms. The organisms are sensitive to minor changes in temperature and light exposure. Antibiotics, antihistamines, phenothiazines and possibly other drugs frequently contained in the patients' sera suppress growth and produce falsely low test results. Further, only nonturbid sera can be examined excluding hyperlipimic or stored sera. The development of radioisotopic techniques has provided an alternative and better assay of B_{12} concentration in blood. The relative simplicity of the procedure and the recent availability of the reagents in kit form has made rapid measurement of serum B_{12} concentration feasible in most community hospitals and medical centers.

TECHNIQUE

All of the assays utilizing radioisotopes for determining of B_{12} concentrations are based on the principle of competitive protein binding in which labeled and unlabeled molecules compete for a constant number of protein binding sites. The princi-

$$1)\qquad IF + B_{12} \underset{k_2}{\overset{k_1}{\rightleftharpoons}} IF\!:\!B_{12}$$

$$2)\qquad IF + B_{12}\!:\!B_{12}^{*} \rightleftharpoons B_{12}\!:\!IF\!:\!B_{12}^{*}$$

Figure 6–1. Principle of competitive protein binding. (1) B_{12} binds to intrinsic factor (IF) according to the law of mass action with the reaction constants k_1 and k_2. (2) In the assay system, labeled and unlabeled B_{12} molecules compete for a limited number of binding sites on intrinsic factor. The more unlabeled B_{12} present the higher the ratio of free to protein-bound B_{12}^{*}; the less present, the lower the ratio.

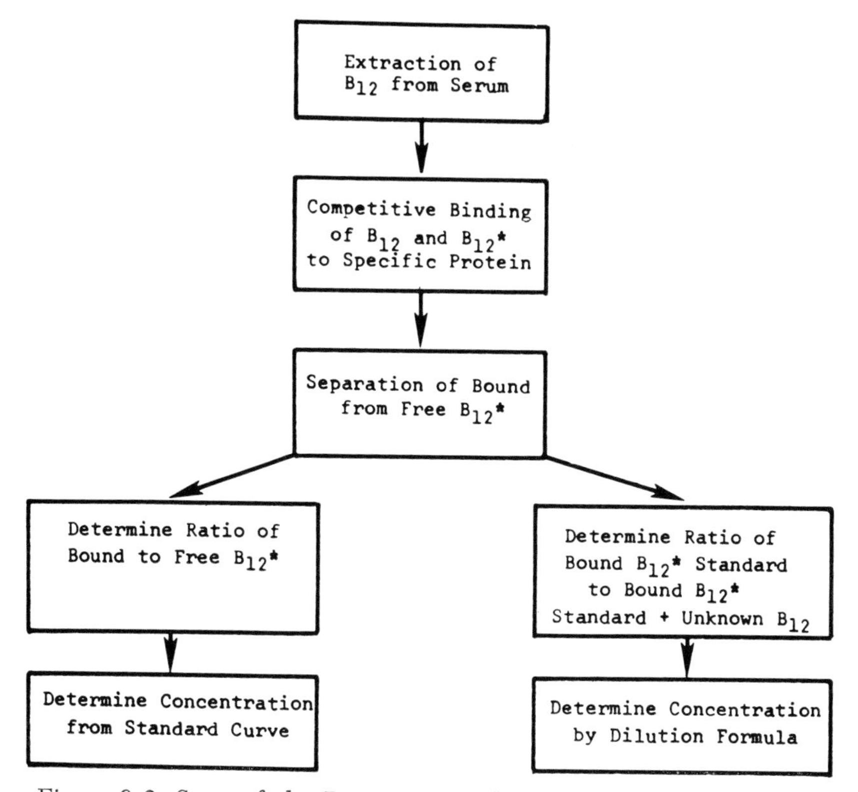

Figure 6–2. Steps of the B_{12} assay procedure common to all radioassay methods.

ple is illustrated in Figure 6–1. Both reactions shown follow the law of mass action, and are allowed to progress to equilibrium or near-equilibrium. The equilibrium is governed by the reaction constants k_1, k_2, k_3 and k_4. The more unlabeled B_{12} present, the higher the ratio of free to protein-bound radioactively labelled B-12 (B_{12}^*); the less present, the lower the ratio. The principle steps of the assay are (Fig. 6–2):

1. Extraction of B_{12} from the sample and destruction of the serum binding proteins.
2. Binding of the endogenous B_{12} and B_{12}^* to the specific assay protein, establishing equilibrium between free B_{12}^*, free B_{12}, B_{12}^* protein complex, and B_{12} protein complex.
3. Separation of free from bound B_{12}^*.

TABLE 6–I
RADIOASSAY METHODS FOR VITAMIN B_{12} CONCENTRATION

INVESTIGATOR	*NORMAL RANGE IN SERUM (pg/ml)*	*EXTRACTION B_{12}*	*ASSAY PROTEIN*	*SEPARATION OF FREE FROM BOUND B_{12}*
Barakat and Ekins [8]	330–840	Boiling at pH 4.5 concentration by evaporation of solvent	Plasma B_{12} binding protein	Dialysis
Rothenberg [9]	160–800	Double boiling at pH 4.6 & 5.6 in cyanide	Intrinsic Factor	Protein precipitation with BaOH-Zn SO_4
Lau et al.[12]	200–900	Heating at pH 1–2	Intrinsic Factor	Coated charcoal
Frenkel et al.[10]	160–800	Double boiling at pH 4.5 & 5.6 in cyanide	Serum B_{12} binding protein	DEAE cellulose
Friedner et al.[11]	160–865	Boiling at pH 4.75–5.0 in cyanide	Intrinsic Factor	Ultrafiltration
Carmel and Coltman [16]	205–937	Heating at pH 1–2	Saliva B_{12} binder	Coated charcoal
Wide and Killander [13]	237–1190	Boiling at pH 3.6, cyanide	Intrinsic Factor-Sephadex Complex	Intrinsic factor-Sephadex complex

4. Determination of the ratio of free to bound B_{12}*.
5. Conversion of the ratio to the concentration of sample B_{12} with a standard curve.

Many assays have been described with the above basic steps in common but with differing techniques. Representative methods of each major modification are listed in Table 6–I. Their large number indicates that no assay method is completely satisfactory.

Extraction of B_{12} from Sample

Vitamin B_{12} is present in blood as hydroxy- and methylocobalamin, which are chemically highly unstable forms. Most of the vitamin circulates bound to specific transport proteins, transcobalamin I and II, glycoproteins with molecular weights of 120,000 and 38,000 respectively.[7] The binding to transport proteins probably renders B_{12} resistant to rapid degradation in the circulation. Several methods of B_{12} extraction have been developed. Most commonly, heating of the serum sample to 100° C at an acid pH (4.5 to 5.6) is used to separate free from bound B_{12}, to precipitate the serum proteins and to destroy their binding ability.[8-11] Herbert's group [12] and Wide et al.[13] acidified the sera to as low as pH 1 and heated gently without causing overt protein precipitation. Other investigators [14-16] have claimed autoclaving of the sample is necessary to destroy nonspecific B_{12} binders that occur in some sera and interfere with the assay. These nonspecific binders were detected particularly in sera which had been stored for several months. Complete removal of serum proteins and all other potential binders is mandatory for assay procedures that utilize competitive binding by assay proteins.[17]

Sodium or potassium cyanide added to the serum converts the vitamin to the cyanocobalamin, an unphysiologic but heat-stable form. It is further believed that the conversion to cyanocobalamin facilitates complete separation of B_{12} from nonspecific binding proteins.[9,15,18] Omission of cyanide in some radioassays probably accounts for the very low B_{12} values found in deficient sera compared to the values obtained by the microbiologic assays where cyanide is routinely used.[11,15]

Assay Proteins

After extraction of the sample B_{12}, $^{57}Co\text{-}B_{12}$ or $^{60}Co\text{-}B_{12}$ is added. $^{57}Co\text{-}B_{12}$ is preferred because of its 121 kev gamma emission which is suitable for radioassay in a well scintillation counter and its availability with a high specific activity (50 to 300 μCi/μg). The picogram quantity of labeled cobalamin added should be approximately equal or preferably less than the smallest amount of vitamin to be detected in the system.[11] The concentration of the protein is adjusted for the assay so that it is 60 to 80 percent saturated by the labeled cobalamin. The optimal concentration of the assay protein is determined by adding the picogram quantity of labeled cobalamin to be used in the assay to increasing dilutions of assay protein thereby determining the concentration at which approximately 70 percent of the labeled cobalamin is bound.[9] The binding qualities of each lot of assay protein should be determined by this procedure because of variability between lots, and loss of binding through prolonged storage.[9,11]

Intrinsic Factor

Because of its high specificity and high affinity for B_{12}, intrinsic factor prepared from hog gastric mucosa has been used in many laboratories.[9,11,12,15] Intrinsic factor offers rapid binding with only slight subsequent dissociation of the B_{12} complex.[19] However, there are a few problems with intrinsic factor binding which can be controlled by rigid technique. The affinity of intrinsic factor varies considerably between different suppliers and different lots of the same supplier. The more highly purified preparations appear less stable when frozen in low concentrations than crude preparations.[9] Several investigators have reported that intrinsic factor binds B_{12} more efficiently in the presence of serum or albumin even at very low concentrations though they offer no satisfactory explanation for this finding.[15,16,20,21] Rothenberg [20] studied the effect of storage of intrinsic factor on the recovery of crystalline B_{12} from test solutions and reported better recovery when serum was used as a diluent rather than saline. Raven [15] confirmed the finding and demonstrated better recovery using B_{12} deficient serum as a diluent. A partial explanation of this phenomenon is probably the loss of intrinsic factor by adherence to glass in protein-poor solution, similar to the problems encountered with the storage of peptide hormones and other proteins. Instability of intrinsic factor in very low concentrations and its variability between different lots makes frequent examination of its binding qualities and the routine use of a standard curve mandatory.

Another important characteristic of intrinsic factor complicating the assay procedure is that its B_{12} binding capacity is not constant. It increases as the total concentration of B_{12} within the system is increased.[21] The B_{12} in samples with low concentrations may be underestimated unless several standards are measured specifically in this range.

Recently, Wide and Killander[13] described an assay utilizing intrinsic factor bound to cross-linked dextran or microcrystalline cellulose. The insoluble polysacchride is treated with cyanobromide (CNBr) to synthesize an intrinsic factor-polysaccharide complex that retains the B_{12} binding ability of intrinsic factor. The insoluble complex is separated by centrifugation from free B_{12} which remains in solution. This obviates the need for a separate binding agent to remove the free B_{12}. A serum B_{12} assay in kit form * utilizing this principle has been marketed with Sephadex-intrinsic factor complex as the B_{12} absorbent.

Serum Binding Proteins

Barakat and Ekins[8] and Frenkel et al.[10] used standard or pooled serum as a source of the assay protein for B_{12} with satisfactory reproducibility on comparison to the microbiologic assay. More recent investigators have defined the serum binders as natural transport proteins: an alpha globulin (transcobalamin I) and a beta globulin (transcobalamin II). Labeled B_{12} added to normal serum *in vitro* will bind predominantly to transcobalamin I but also, though less firmly, to transcobalamin II. To avoid the complications introduced by two different classes of binding sites

* Phadebas Vitamin B_{12} test. Pharmacia Laboratories, Inc., Piscataway, New Jersey.

as found in normal human serum, Rothenberg [21] used serum from a patient with chronic myelogenous leukemia which contained large amounts of transcobalamin I but very little transcobalamin II. He demonstrated that transcobalamin I did not exhibit as great an increase of B_{12} binding capacity with increasing concentrations of B_{12} as did intrinsic factor and normal serum. The binding capacity of the transcobalamin I did not change with storage, after freezing and thawing, even after dilution to the low concentrations to be used in the assay. Transcobalamin I obtained from leukemic blood proved to be a better assay protein than transcobalamins I and II combined, as obtained from normal serum.

In procedures utilizing plasma B_{12} binding proteins, incubation with B_{12} must be carried out for at least one hour. Shorter incubation time does not permit maximal binding and affects the reproducibility of the assay adversely.[10,18]

Human Saliva Protein

Human saliva contains a single B_{12} binding protein with a molecular weight of 60,000 resembling transcobalamin I in its immunologic and electrophoretic properties. Carmel and Coltman [16] found the B_{12} binding capacity of saliva from seven normal subjects to average 40 ng/ml. The assay protein, diluted in phosphate buffer to a concentration suitable for B_{12} assay, was stable in the frozen state for as long as four months. Maximal binding of the B_{12} to the salivary protein was achieved within thirty minutes. Increasing B_{12} concentration did not cause increased binding in contrast to intrinsic factor and transcobalamin I and II in normal sera. Contamination with serum, however, increases B_{12} binding by saliva causing overestimation of B_{12} in the sample. Construction of a standard curve with serum added to the reaction mixture corrects for this problem and yields useful results.

Separation of Free from Bound B_{12}

Of the many methods reported to separate the free B_{12} from the B_{12} protein complex the most attractive are those utilizing

DEAE cellulose or coated charcoal. Earlier methods are noted for reasons of completeness.

Dialysis

Barakat and Ekins,[8] pioneers in the field of radioassay of hormones and vitamins, tried dialysis of serum against a sixty-fold volume of saline. Dialysis was carried out for forty-eight hours, presumably to achieve equilibrium or another reproducible endpoint. No information is given in regard to temperature requirements, bacterial growth and possible degradation of labeled B_{12} during dialysis. It appears reasonable to assume that the investigators abandoned this approach in favor of more promising technique.

Ultrafiltration by High Speed Centrifugation

Friedner et al.[11] used ultrafiltration by high speed centrifugation to separate free from bound radiolabeled B_{12}. They placed B_{12} equilibrated serum into moistened dialysis membrane bags suspended in empty tubes and centrifuged for seventy-five minutes at 1,000 xg at less than 15° C. The radioactivity of 0.5 ml of ultrafiltrate (the free B_{12}) was compared to a standard curve obtained in the same manner. A serious disadvantage of this procedure was dilution of the free B_{12} in the ultrafiltrate by moisture from the wall of the dialysis membrane. This dilution appeared to be constant and acceptable results were obtained if standards were run at the same time. Careful attention to the time and speed of centrifugation was required for reproducible results.

Protein Precipitation with Barium Hydroxide-Zinc Sulfate

Rothenberg, using intrinsic factor[9] or transcobalamin I[21] as the assay protein achieved separation by precipitation of the B_{12} protein complex with barium hydroxide and zinc sulfate. Complete separation of the intrinsic factor-B_{12}^* from B_{12}^* was facilitated by addition of human serum albumin. The pH was kept below 8 and monitored with phenolphthalein. This method, while attractive, has not gained wide acceptance.

Absorption by Coated Charcoal

Miller[22] reported that free B_{12} is absorbed by activated charcoal while protein-bound B_{12} is not bound. Several investigators made use of this finding by designing B_{12} assay procedures.[12,14-16] Gottlieb, et al.[23] demonstrated that selective absorption of free B_{12} was achieved only when the charcoal was coated with large molecules such as serum proteins, which blocked absorption of other proteins such as intrinsic factor, but allowed absorption of small molecules such as free B_{12}. Herbert's group[12] found that albumin-coated charcoal occasionally failed to bind up to 5 percent of the free B_{12} while binding by hemoglobin-coated charcoal was virtually complete. Carmel and Coltman[16] reported that some brands of activated charcoal, when coated with hemoglobin, absorbed B_{12}-intrinsic factor complex leading to a falsely low result in the assay. The problem was corrected by re-coating the charcoal with serum proteins. It appears advisable to examine the binding specificity of protein-coated charcoal before use.

Absorption by DEAE Cellulose

B_{12} protein complexes bind firmly to diethyl-aminoethyl (DEAE) cellulose in 0.01 M phosphate buffer at an alkaline pH.[24] In the assay of Frenkel et al.,[10] a B_{12}-transcobalamin I and II-DEAE cellulose complex is formed which is insoluble and easily separated from the reaction mixture by centrifugation. Care must be taken that sufficient cellulose is added to quantitatively absorb all the B_{12}-protein complex.[18]

Calculation of the B_{12} Concentration

In most of the assay procedures described above, the quantity of B_{12} in a given sample is determined by reference to a standard curve obtained from standards processed simultaneously with the unknown. Five to seven, or preferably more, crystalline B_{12} standards are selected appropriate to the anticipated concentration of B_{12} in the sample to be processed. In methods utilizing B_{12} binders with altered binding capacity at low concentrations of B_{12}, such as intrinsic factor and transcobalamin I and II,[21] at least two or three standards should be in this range. The distribution of the labeled B_{12}* between free and bound fractions can be expressed and illustrated in several ways. Most commonly, the ratio of bound to free B_{12}* is plotted against the standards on a linear scale (Fig. 6–3). For wider ranges a logarithmic scale is more convenient and has the added advantage of transforming a major portion of the standard curve into a straight line, facilitating data conversion (Fig. 6–4).

In an effort to simplify the assay procedure, Lau et al.[12] eliminated the use of a standard curve. They used a single control instead, which contained no unlabeled B_{12}. An excess of labeled B_{12} sufficient in quantity to saturate all binding sites on the assay protein was added to sample and control. The free B_{12} was removed by coated charcoal absorption. The activity of the radio-B_{12} bound to intrinsic factor in the supernatant was used as a standard. Displacement of the labeled vitamin by the non-labeled vitamin in the serum sample was assumed to be proportional to the concentration of the latter which could be determined by the dilution formula: [25]

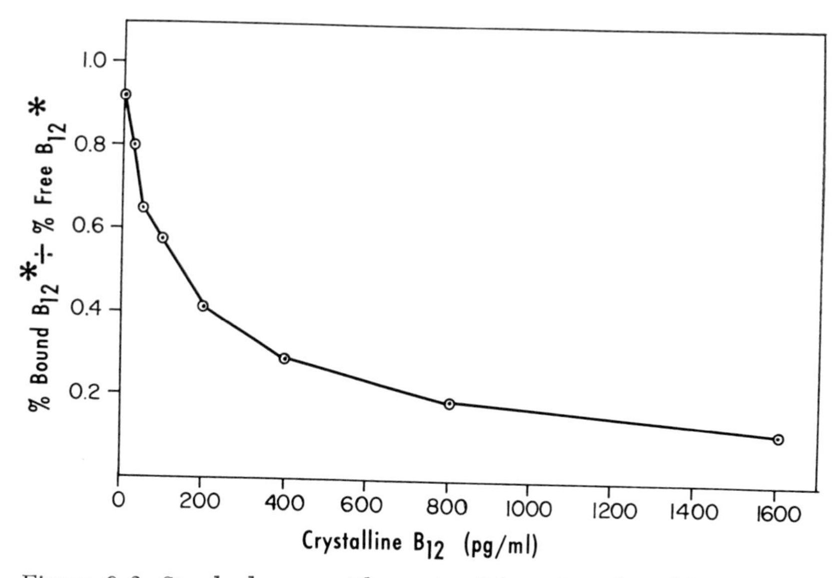

Figure 6–3. Standard curve. The ratio of bound to free labeled B_{12} is plotted for each of the crystalline B_{12} standards on a linear scale.

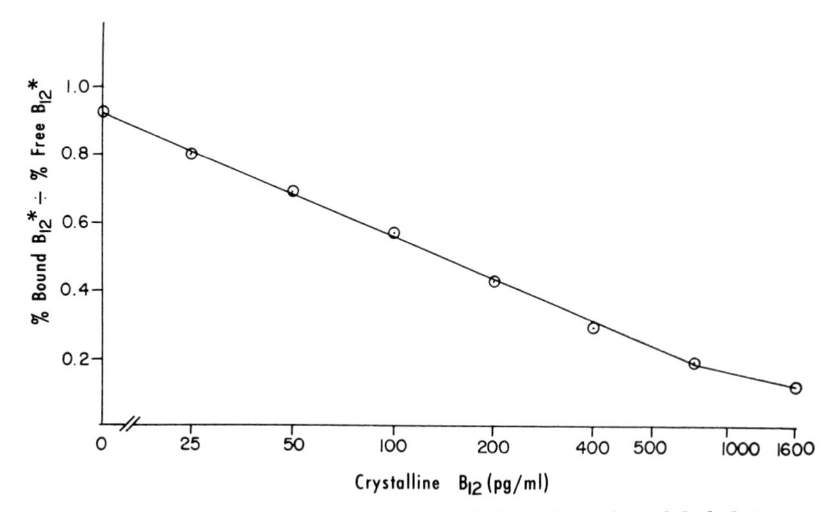

Figure 6–4. Standard curve. The ratio of bound to free labeled B_{12} is plotted for each of the crystalline B_{12} standards on a logarithmic scale.

$$\text{Sample } B_{12}\text{ (pg)} = {}^{57}\text{CO } B_{12}\text{ (pg)} \times \left(\frac{B}{B_1} - 1\right),$$

where B = net counts of standard bound to intrinsic factor, and B_1 = net counts of standard and nonlabeled B_{12} bound to intrinsic factor.

Lau and associates [12] claimed good correlation of the results obtained by their method with those obtained with techniques using a standard curve. Raven et al.[15] obtained good correlation of a single standard control radioassay method with the microbiologic assay, only when B_{12}-deficient serum was used as a diluent. The reliability of this method is dependent upon accurate determination of the specific activity of labeled B_{12} as supplied by the commercial producer. Though the simplicity of the assay is tempting, caution is indicated until its reliability is fully established for the entire range of expected B_{12} concentrations.

COMPARISON OF RADIOASSAY AND MICROBIOLOGIC ASSAY

Most investigators compared their radioassay methods to the microbiologic assay and found a satisfactory correlation, particularly when a standard curve was used.[10,11,15,18] However, Lau et al.[12] and Raven et al.,[15] using coated charcoal as an absorbent, found lower levels than with the microbiologic assay. Subsequently, Raven et al.[26] and Liu and Sullivan,[27] using a similar technique, discovered the reason for this discrepancy. Apparently conversion of hydroxy- and methylocobalamine to the more stable form, cyanocobalamine, prevents degradation of the vitamin during bioassay. It is conceivable that this phenomenon is peculiar to this assay.

Wide and Killander,[13] using a solid phase, radiosorbent technique, reported good correlation with a microbiologic assay using *Euglena gracilis*. However, the B_{12} concentrations obtained by the radiosorbent technique were numerically higher. The normal range for their assay is higher than reported by other investigators (Table 6–I). No explanation is offered.

CLINICAL APPLICATION

Normal B_{12} Levels

Examination of the sera from large groups of normal subjects has demonstrated that serum B_{12} levels has a non-Gaussian distribution (Fig. 6–5) with the mean of 400 to 500 pg/ml and a lower limit at 160 pg/ml.[28,29] As shown in Table 6–I, the normal range of serum B_{12} as determined by four of the assays is 160 to 850 pg/ml. The range found with the three other assays is somewhat higher. While it has not been determined which of the assays is the most accurate, each assay provides the same clinically useful information.

With increasing age there is a trend toward lower serum B_{12} levels.[28] Serum B_{12} levels also decline progressively during pregnancy in association with an increase in B_{12} binding proteins.[30-32] Depressed serum B_{12} levels have also been reported in normal women receiving oral contraceptives of a combination type.[33] As there was no impairment of B_{12} absorption or hematologic aberrations in these patients, the cause of this phenomen is also an alteration in B_{12} binding proteins.

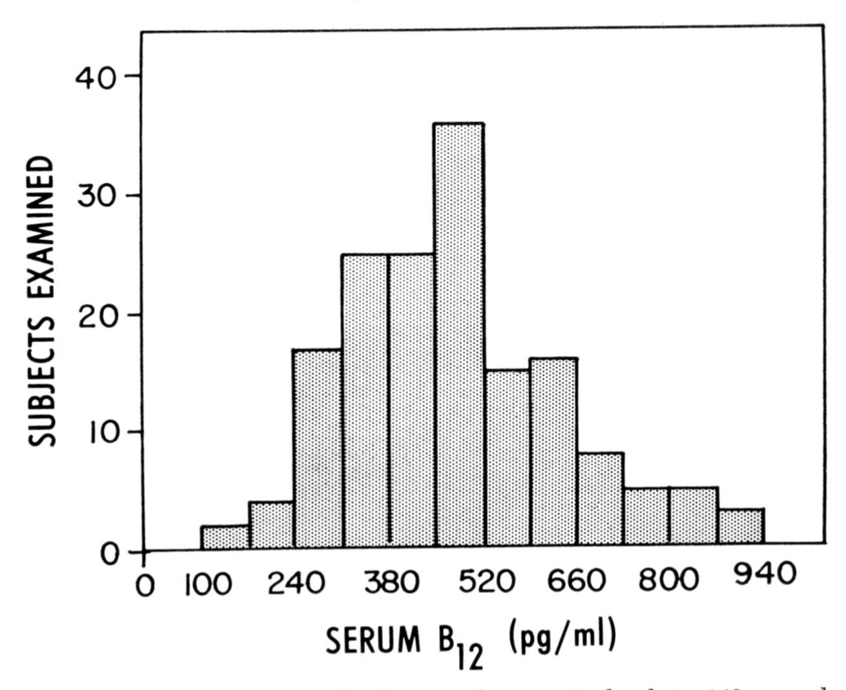

Figure 6–5. The frequency distribution of serum B_{12} levels in 149 normal subjects, ages twenty to fifty, is shown. The mean serum B_{12} level was 472 pg/ml. The distribution is non-Gaussian with a skew to lower values. (Modified from B.B. Anderson, *J Clin Pathol, 17*:14, 1964.)

Disorders with Low Serum B_{12} Levels

Analysis of the serum B_{12} level has proved to be an efficient screening technique for the detection of B_{12} deficiency. Many investigators have demonstrated good separation between the low serum B_{12} levels in patients with B_{12} responsive megaloblastic anemia, and the levels in normal subjects.[8,9,16,21,29,31] In a series of patients with pernicious anemia who were examined monthly after B_{12} therapy was interrupted, megaloblastosis occurred when the serum B_{12} level was between 70 and 154 pg/ml.[34] Patients with megaloblastic anemia secondary to dietary folic acid deficiency have normal or elevated serum B_{12} levels.[10,27,31] If folic acid is administered to patients with B_{12} deficiency, the serum B_{12} level may decrease even further.[35] Administration of folic acid to prevent folic acid deficiency in patients on anticonvulsant therapy is associated with depression of serum B_{12} levels.[36]

Prospective and retrospective studies of patients following subtotal gastrectomy demonstrate that the serum B_{12} concentration may be low before the bone marrow or peripheral blood manifestation of B_{12} deficiency become evident.[37-39] Likewise, patients with subacute combined degeneration of the spinal cord secondary to pernicious anemia have been reported with abnormally low serum B_{12} concentrations without significant anemia.[40] In this regard, serum B_{12} concentration is a useful tool

TABLE 6–II

CLINICAL STATES ASSOCIATED WITH DECREASED SERUM B_{12} CONCENTRATION

B_{12} Deficient States:

Pernicious Anemia: Addisonian or juvenile.
Combined system degeneration, with or without anemia.
Small intestinal disease: regional enteritis, Whipple's Disease, tropical and nontropical sprue.
Ileal resection or bypass.
Intestinal blind loops with bacterial overgrowth.
Fish tapeworm infestation.
Para-aminosalicylic acid administration.
Complete dietary lack of animal food products.

Alterations of Plasma Binding Proteins:

Pregnancy.
Oral contraceptive administration.
Vitamin B_{12} binding alpha globulin deficiency.

to separate combined system degeneration without hematologic manifestations from intrinsic spinal cord disease.

It is well accepted that vitamin B_{12} deficiency is produced by four principal mechanisms: dietary deficiencies, intrinsic factor deficiency, intestinal malabsorption, or competition for B_{12} by intestinal organisms. Dietary deficiency of vitamin B_{12} is rare in the United States, but is seen in countries where animal protein, the main dietary source of B_{12} is rarely consumed.

Intrinsic factor deficiency is the defect responsible for B_{12} malabsorption in classic pernicious anemia and in a specific type of juvenile pernicious anemia. Intrinsic factor deficiency is also a frequent sequel to total or subtotal gastrectomy; its incidence increasing with time following the surgery, eventually resulting in megaloblastic anemia.[37,39] Screening of postgastrectomy patients has revealed that up to 75 percent have decreased serum B_{12} levels, while tests which measure the ability to absorb B_{12} may remain normal.[41] In this clinical setting, determination of serum B_{12} concentration is clearly superior to the absorption tests.

Small intestinal diseases including regional ileitis, Whipple's disease, tropical and nontropical sprue, as well as surgical bypass or resection of the ileum may produce an inability to absorb B_{12} intrinsic factor complex. Administration of colchicine,[42] para-amino salicylic acid,[43] the biguanides,[44] and alcohol[45] can also produce B_{12} malabsorption. While these drugs will produce an abnormal B_{12} absorption test, they would not be expected to produce a low serum B_{12} level unless administration was continued over several years to allow for depletion of body B_{12} stores. Heinivarra and Palva,[43] however, have reported abnormally low serum B_{12} levels in twenty-five of thirty-two patients who had been taking para-amino salicylic acid for little more than one year. None of these patients had developed megaloblastic bone marrow.

Impaired B_{12} absorption and decreased serum B_{12} levels have been found in numerous patients with various anatomic lesions of the small intestine, including surgically produced blind loops, in which colonic bacteria proliferate in the small intestine. The failure to absorb B_{12} appears due to competition for B_{12} by the microorganisms.[46,47] Suppression of the bacterial growth with antibiotics restores normal B_{12} absorption in this situation.

Infestation with the fish tapeworm, *Diphyllobothrium latum*, also produces B_{12} malabsorption apparently due to competition for B_{12} between host and parasite. B_{12} depletion is rarely very prominent; only 3 percent of infested persons become anemic. While both the infested anemic patients and the nonanemic infested patients have low serum B_{12} concentrations, those patients who were anemic had significantly lower levels.[48] Severe B_{12} deficiency states due to fish tapeworm infestation are rarely found outside eastern Finland.

While folic acid deficiency is the most common cause of megaloblastosis in childhood, infants of B_{12} deficient mothers may demonstrate megaloblastosis and abnormally low serum B_{12} levels.[49]

Abnormally low serum B_{12} concentration without apparent B_{12} deficiency has been reported in two brothers in association with a deficiency of the Alpha globulin B_{12} binder. There was no megaloblastosis and only a transient rise in serum B_{12} level was observed following parenteral B_{12} administration.[50] This state is analogous to hereditary absence of other binding proteins such as thyroid binding globulin deficiency.

DISORDERS WITH HIGH SERUM B_{12} LEVELS

Increased serum B_{12} levels have been found in up to 36 percent of patients with untreated polycythemia rubra vera, in myelofibrosis with myeloid metaplasia and in chronic myelogenous leukemia.[16,32,51,52] In these conditions, B_{12} levels are more often normal than elevated. Following control of polycythemia by myelosuppressive drugs, the B_{12} concentration may return to normal level but this phenomenon is not observed following phlebotomy.[51] Increased serum B_{12} concentration has also been observed in acute myelocytic leukemia and has been suggested as a means of differentiating this form of leukemia from acute lymphoblastic leukemia.[53] The increased serum B_{12} in the myeloproliferative disorders is associated with, and probably secondary to, an increase in serum transcobalamin I or a similarly acting plasma binder.[32,54] Similar aberrations have been described in patients with a leukemoid reaction secondary to acute infec-

TABLE 6–III
CLINICAL STATES ASSOCIATED WITH ELEVATED SERUM B_{12} CONCENTRATION

Polycythemia Rubra Vera
Chronic Myelogenous Leukemia
Myelofibrosis with Myeloid Metaplasia
Acute Myelogenous Leukemia
Leukemoid Reaction
Viral Hepatitis
Alcoholic Cirrhosis
Acute Renal Failure

tion.[55] Pernicious anemia with a serum B_{12} concentration in the normal range (228 pg/ml) has been reported in a patient with chronic myelogenous leukemia.[56] Following B_{12} administration, the serum B_{12} concentration rose to greater than 1600 pg/ml. It was suggested that circulating B_{12} was tightly bound to the abnormal binding protein and not available to tissues resulting in megaloblastosis.

Elevated serum B_{12} levels have also been found in patients with renal failure,[57] and in patients with liver disease with hepatocellular damage, particularly acute viral hepatitis.[58,59] High B_{12} levels support the established diagnosis of the listed disorders, and may be of differential diagnostic value.

SUMMARY

The radioassay technique for vitamin B_{12} level has definite advantages over microbiologic assay methods: simplicity, speed and reproducibility. It also allows examination of turbid or hyperlipemic sera. The radioassay correlates well with the microbiologic assay. Serum B_{12} determination is a very useful clinical tool in the diagnosis of vitamin B_{12} deficient state. Serum B_{12} level determination is an excellent screening test for the many conditions associated with B_{12} deficiency and B_{12} malabsorption.

REFERENCES

1. Rickes, E.L., Brink, N.G., Koniuszy, F.R., Wood, T.R., and Folkers, K.: Crystalline vitamin B_{12}. *Science, 107:*396, 1948.
2. Berk, L., Castle, W.B., Welch, A.P., Heinle, R.W., Anker, R., and Epstein, M.: Observations on the etiologic relationship of achylia gastrica to pernicious anemia. X Activity of vitamin B_{12} as food (extrinsic) factor. *N Engl J Med, 239:*911, 1948.
3. Gardner, F.H., Harris, J.W., Schilling, R.F., and Castle, W.B.: Observations on the etiologic relationship of achylia gastria to pernicious anemia. XI Hematopoietic activity in pernicious anemia of beef muscle extract containing food (extrinsic) factor upon intravenous injection without contact with gastric (intrinsic) factor. *J Lab Clin Med., 34:*1502, 1949.
4. Schilling, R.F.: Intrinsic factor studies II. The effect of gastric juice on the urinary excretion of radioactivity after the oral administration of radioactive vitamin B_{12}. *J Lab Clin Med, 42:*860, 1953.

5. Hutner, S.H., Bach, M.K., and Ross, G.I.M.: A sugar-containing basal medium for vitamin B_{12} assay with *Euglena;* application to body fluids. *J Protozool, 3:*101, 1956.
6. Spray, G.H.: An improved method for the rapid estimation of vitamin B_{12} in serum. *Clin Sci, 14:*661, 1955.
7. Hall, C.A.: Vitamin B_{12}-binding proteins of man. *Ann Intern Med, 75:*297, 1971.
8. Barakat, R.M., and Ekins, R.P.: Assay of vitamin B_{12} in blood: a simple method. *Lancet, 2:*25, 1961.
9. Rothenberg, S.P.: Radioassay of serum vitamin B_{12} by quantitating the competition between ^{57}Co B_{12} and unlabeled B_{12} for binding sites of intrinsic factor. *J Clin Invest, 42:*1391, 1963.
10. Frenkel, E.P., Keller, S., and McCall, M.S.: Radioisotopic assay of serum vitamin B_{12} with the use of DEAE cellulose. *J Lab Clin Med, 68:*510, 1966.
11. Friedner, S., Josephson, B., and Levin, K.: Vitamin B_{12} determination by means of radioisotope dilution and ultrafiltration. *Clin Chim Acta, 24:*171, 1969.
12. Lau, K.S., Gottlieb, C.W., Wasserman, L. R., and Herbert, V.: Measurement of serum vitamin B_{12} level using radioisotope dilution and coated charcoal. *Blood, 26:*202, 1965.
13. Wide, L., and Killander, A.: A radiosorbent technique for the assay of serum vitamin B_{12}. *Scand J Clin Lab Invest, 27:*151, 1971.
14. Grossowicz, D.D., Sulitzeanu, D., and Merzbach, D.: Isotopic determination of vitamin B_{12} binding capacity and concentration. *Proc Soc Exp Biol Med, 109:*604, 1962.
15. Raven, J.L., Robson, M.B., Walker, R.L., and Barkhan, P.: The effect of cyanide, serum and other factors on the assay of vitamin B_{12} by a radioisotope method using ^{57}Co B_{12}, intrinsic factor and coated charcoal. *Guys Hosp Rep, 117:*89, 1968.
16. Carmel, R., and Coltman, C.A.: Radioassay for serum vitamin B_{12} with the use of saliva as the vitamin B_{12} binder. *J Lab Clin Med, 74:*967, 1969.
17. Murphy, B.E.P.: Protein binding and the assay of nonantigenic hormones. *Recent Prog Horm Res, 25:*563, 1969.
18. Tibbling, G.: A method for determination of vitamin B_{12} in serum by radioassay. *Clin Chim Acta, 23:*209, 1969.
19. Bunge, M.B., and Schilling, R.F.: Intrinsic factor studies. VI. Competition for vitamin B_{12} binding sites offered by analogues of the vitamin. *Proc Soc Exp Biol Med., 96:*587, 1957.
20. Rothenberg, S.P.: Assay of serum vitamin B_{12} concentration using ^{57}Co B_{12} and intrinsic factor. *Proc Soc Exp Biol Med, 108:*45, 1961.
21. Rothenberg, S.P.: A radioassay for serum B_{12} using unsaturated transcobalamine I as the B_{12} binding protein. *Blood, 31:*44, 1968.

22. Miller, O.N.: Determination of bound vitamin B_{12}. *Arch Biochem Biophys, 68:*255, 1957.
23. Gottlieb, C., Lau, K., Wasserman, L., and Herbert, V.: Rapid charcoal assay for intrinsic factor (IF) gastric juice unsaturated B_{12} binding capacity, antibody to IF, and serum unsaturated B_{12} binding capacity. *Blood, 25:*875, 1965.
24. Fahey, J.L., McCoy, P.F., and Goulian, M.: Chromatography of serum proteins in normal and pathologic sera: the distribution of protein-bound carbohydrate and cholesterol, thyroxine-binding protein, B_{12} binding protein, alkaline and acid phosphases, radioiodinated albumin and myeloma proteins. *J Clin Invest, 37:*272, 1958.
25. Rittenberg, D., and Foster, G.L.: A new procedure for quantitative analysis by isotope dilution with application to the determination of amino acids and fatty acids. *J Biol Chem, 133:*737, 1940.
26. Raven, J.L., Robson, M.B., Walker, P.L., and Barkham, P.: Improved method for measuring vitamin B_{12} in serum using intrinsic factor, ^{57}Co B_{12}, and coated charcoal. *J Clin Pathol, 22:*205, 1969.
27. Liu, J.K., and Sullivan, L.W.: An improved radioisotope dilution assay for serum vitamin B_{12} using Hemoglobin-coated charcoal. *Blood, 39:*426, 1972.
28. Boger, W.P., Wright, L.N., Strickland, J.C., Gylfe, J.S., and Ciminera, J.L.: Vitamin B_{12}: correlation of serum concentrations and age. *Proc Soc Exp Bio Med, 89:*375, 1955.
29. Anderson, B.B.: Investigations into the *Euglena* method for the assay of the vitamin B_{12} in serum. *J Clin Pathol, 17:*14, 1964.
30. Boger, W.P., Wright, L.D., Beck, G.D., and Bayne, G.M.: Vitamin B_{12}: correlation of serum concentrations and pregnancy. *Proc Soc Exp Biol Med, 92:*140–143, 1956.
31. Spray, G.H., and Witts, L.J.: Results of three years' experience with microbiological assay of vitamin B_{12} in serum. *Br Med J, 1:*295, 1958.
32. Herbert, V.: Diagnostic and prognostic values of measurement of serum vitamin B_{12}-binding proteins. *Blood, 32:*305, 1968.
33. Wertalik, L.F., Metz, E.N., LoBuglio, A.F., and Balcerzak, S.P.: Decreased serum B_{12} levels with oral contraceptive use. *JAMA, 221:*1371, 1972.
34. Adams, F., Boddy, K., and Douglas, A.S.: Interrelationship of serum vitamin B_{12}, total body vitamin B_{12}, peripheral blood morphology, and the nature of erythropoiesis. *Br J Haematol, 23:*297, 1972.
35. Bok, J., Faber, J.G., DeVries, J.A., Kroese, W.F.S., and Nieweg, H.O.: The effect of pteroylglutamic acid administration on the serum vitamin B_{12} concentration in pernicious anemia in relapse. *J Lab Clin Med, 51:*667, 1958.
36. Hunter, R., Barnes, J., and Matthews, D.M.: Effect of folic acid supplement on serum-vitamin B_{12} levels in patients on anticonvulsants. *Lancet, 2:*309, 1969.

37. Deller, D.J., and Witts, L.J.: Change in the blood after partial gastrectomy with special reference to vitamin B_{12} I. Serum vitamin B_{12}, haemoglobin, serum iron, and bone marrow. *Q J Med, 31:*71, 1962.
38. Ungar, B., and Cowling, D.C.: Vitamin B_{12} studies after gastrectomy. *Med J Aust, 2:*861, 1962.
39. Hines, J.D., Hoffbrand, A.V., and Mollin, D.L.: The hematologic complications following partial gastrectomy: a study of 292 patients. *Am J Med, 45:*555, 1967.
40. Victor, M., and Lear, A.A.: Subacute combined degeneration of the spinal cord. *Am J Med, 20:*896, 1956.
41. Mahmud, K., Ripley, D., and Doscherholmen, A.: Vitamin B_{12} absorption tests. Their unreliability in past gastrectomy states. *JAMA, 216:*1167, 1971.
42. Webb, D.I., Chodos, R.B., Mahar, C.Q., and Faloon, W.W.: Mechanism of vitamin B_{12} malabsorption in patients receiving colchicine. *N Engl J Med, 279:*845, 1968.
43. Heinivaara, O., and Palva, I.P.: Malabsorption and deficiency of vitamin B_{12} caused by treatment with para-aminosolicylic acid. *Acta Med Scand, 177:*337, 1965.
44. Herbert, V.: Metformin and B_{12} malabsorption. *Ann Intern Med, 76:*140, 1972.
45. Lindenbaum, J., and Lieber, C.S.: Alcohol-induced malabsorption of vitamin B_{12} in man. *Nature, 224:*806, 1969.
46. Halstead, J.A., Lewis, P.M., and Gasster, M.: Absorption of radioactive vitamin B_{12} in the syndrome of megaloblastic anemias associated with intestinal stricture or anastomosis. *Am J Med, 20:*42, 1956.
47. Dellipiani, A.W., Samson, R.R., and Girdwood, R.H.: The uptake of vitamin B_{12} by *E. Coli:* possible significance in relation to the blind loop syndrome. *Am J Dig Dis, 13:*718, 1968.
48. Nyberg, W.: *Diphylobothrium Latum* and human nutrition with particular reference to vitamin B_{12} deficiency. *Proc Nutr Soc, 22:*8, 1963.
49. Jadhau, M., Webb, J.K.G., Vaishnava, S., and Baker, S.J.: Vitamin B_{12} deficiency in Indian infants. *Lancet, 2:*903, 1962.
50. Carmel, R., and Herbert, V.: Deficiency of vitamin B_{12}-binding alpha globulin in two brothers. *Blood, 33:*1, 1969.
51. Gilbert, H.S., Krauss, S., Pasternack, B., Herbert, V., and Wasserman, L.R.: Serum vitamin B_{12} content and unsaturated vitamin B_{12} capacity in myeloproliferative disease. *Ann Intern Med, 71:*719, 1969.
52. Halstead, J.A., Carroll, J., and Rubert, S.: Serum and tissue concentration of vitamin B_{12} in certain pathologic states. *N Engl J Med, 260:*575, 1959.
53. Stahlberg, K.G., Olsson, I., Gahrton, G., and Narden, A.: Serum vitamin B_{12} determination and cytochemical reaction in the differential diagnosis of acute leukemia. *Acta Med Scand, 174:*105, 1963.

54. Hall, C.A., and Finkler, A.E.: Vitamin B_{12} binding protein in polycythemia vera plasma. *J Lab Clin Med, 73:*60, 1969.
55. Carmel, R., and Coltman, C.A.: Nonleukemic elevation of serum vitamin B_{12} and B_{12} binding capacity levels resembling that in chronic myelogenous leukemia. *J Lab Clin Med, 78:*289, 1971.
56. Britt, R.P., and Rose, D.P.: Pernicious anemia with a normal serum vitamin B_{12} level in a case of chronic granulocytic leukemia. *Arch Intern Med, 117:*32, 1966.
57. Matthews, D.M., and Beckett, A.G.: Serum vitamin B_{12} in renal failure. *J Clin Pathol, 15:*456, 1962.
58. Rachmilewitz, M., Aronovitch, J., and Grossowicz, N.: Serum concentrations of vitamin B_{12} in acute and chronic liver disease. *J Lab Clin Med, 48:*339, 1956.
59. Deller, D.J., Kimber, C.L., and Ibbotson, R.N.: Folic and deficiency in cirrhosis of the liver. *Gastroenterology, 10:*35, 1965.

CHAPTER 7

RADIOASSAY OF SERUM TRIIODOTHYRONINE: A NEW CHEMICALLY SPECIFIC METHOD

Garrett A. Hagen and Lincoln L. Diuguid *

INTRODUCTION

For many years following the report of isolation of 1-thyroxine (T4) from the thyroid gland by Kendall in 1919,[1] it was assumed that there was only one biologically active thyroid hormone. However, in 1952, more than thirty years later, Gross and Pitt-Rivers reported the presence of a second thyroid hormone, 3,5,3′ triiodothyronine (T3).[2] Radioactively labeled T3 was demonstrated to be circulating in the blood of patients with thyrotoxicosis who had received large doses of ^{131}I. The newly discovered compound was soon found to be about five times more potent by weight than T4 by means of goiter prevention experiments and effects on persons with myxedema.[3] It was even postulated that T3 might well be the real physiologically active thyroid hormone and that T4 was merely a percursor.[4] Initial evidence suggesting extrathyroidal conversion of circulating T4 to T3 could not be confirmed, however.[5,6] In subsequent years a number of patients were reported in whom elevated serum T3 levels were demonstrated by qualitative methods [7-11] and indeed in some cases T3 appeared to be the sole circulating thyroid hormone.[12-16] In general, however, T4 exceeded the level of T3 by at least twenty to thirty times.[17,18] It was found in humans that T3 produced calorigenic effects sooner and was more rapidly disposed of by the body than T4.[19,20]

* We are grateful to Dr. Thomas F. Frawley for use of the case history of his patient with T3-toxicosis.

However, substantial progress awaited the development of a sensitive quantitative assasy for circulating serum T3. The competitive protein binding method was initially applied to T3 assay with success in 1967 by Nauman et al.[21] and subsequently by Sterling and coworkers.[22] This method has produced a significant amount of preliminary data but is a complex procedure reproduced with difficulty by others. A major problem is elevation of the T3 value due to intraassay deiodination of T4 to form T3.[23,24] Also at this time attempts were made to assay serum T3 by gas chromatography, the most notable success being achieved by Hollander who reported preliminary data from this method.[25] Not surprisingly, these workers all found that serum T3 was depressed in hypothyroidism and elevated in thyrotoxicosis. Initial normal mean levels ranged from 220 to 450 nanograms per 100 ml.

Measurement of serum T3 quickly provided new information concerning its physiological role. New evidence was reported for extrathyroidal conversion of T4 to T3.[26,27] Indeed, pending revision by more precise T3 measurements, 33 percent of total T4 production may be devoted to this pathway, accounting for 41 percent of the T3 produced in the body daily.[28] However, the relative rates of secretion of T3 and T4 by the human thyroid remain to be clarified. T3 has a volume of distribution in the normal human of about 40 liters, roughly four times that of T4. The half-time of circulating serum T3 is about one day compared to one week for T4.[29] An absolute daily disposal rate of 60 micrograms has been reported for T3, a value which approaches the 80 microgram figure for T4.[30] Since T3 is calorigenically about four times more potent than T4, it would appear that most of the metabolic role of the thyroid gland is effected via T3.[29] Although considerable T3 may be derived from extrathyroidal T4, it is undoubtedly also secreted directly by the thyroid gland.

Serum T3 assay has also identified cases of thyrotoxicosis due to elevation of T3 alone, a finding of great interest to the clinician. Initially reported by Hollander, the existence of this entity has subsequently been confirmed by other investigators.[25,31,32]

DOUBLE ISOTOPE DERIVATIVE ASSAY OF SERUM T3

The primary purpose of this paper is to outline a new method for radioassay of circulating T3, applying the double isotope derivative technique. This principle of assay was developed by Udenfriend and others for the measurement of amino acids and subsequently adapted to steroid assays.[33,34] It was successfully used for serum thyroxine assay and reported by Whitehead and Beale in 1959.[35] Double-isotope derivative T3 assay was recently attempted by Sterling et al. and a few determinations reported.[22] We have previously developed and reported a new and different T4 method using the double isotope derivative principle.[36,37] This method has now been adapted for measurement of T3.[38] It has been reported in detail elsewhere,[39] and will be briefly summarized for the purpose of this review.

In the double isotope derivative assay of T3 the unknown parent compound is tagged or labeled by reaction with tritium-

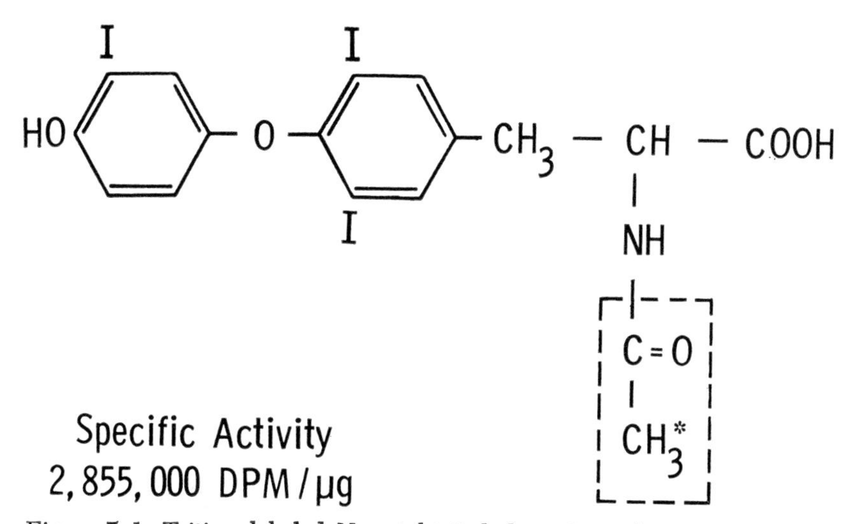

Figure 7–1. Tritium-labeled N-acetyl-triiodothyronine. The experimentally determined specific activity of a derivative prepared from one lot of tritium-labeled acetic anhydride is shown.

labeled acetic anhydride of known specific activity to form an acetylated derivative compound (Fig. 7–1). The labeled derivative is then purified and counted for tritium activity. Since its specific activity is known, the amount of T3 present can be calculated. However, the purification inevitably results in some loss of the compound being assayed. To correct for loss it is necessary to add a tracer amount of T3 labeled with a different isotope at the beginning of the assay. This compound is also counted at the end of the assay and compared with a standard to estimate percent recovery. In this manner the final result can be adjusted for loss occuring during the assay procedure.

T3 Assay Method

Incubation and Extraction

Fifty microliters of ^{131}I T3 solution (10,000 cpm or 0.3 ng) were added to 5 ml of serum in a round-bottomed 50 ml centrifuge tube. The serum and ^{131}I T3 were mixed by gentle agitation at 23° C for forty-five minutes with a Dubnoff Incubator Shaker. Following adjustment to pH 4 by addition of 5.0 ml of 0.2 M acetate buffer, pH 2.6, the sample was incubated overnight (15 hours) with 10 mg crude papain, 0.2 ml 0.1 M EDTA, and 0.2 ml 0.1 M $Na_2S_2O_3$.[40] The digested serum was then applied to an anion exchange resin column by aspiration through a 3 mm diameter polyethylene tube via a peristaltic pump.[37] Twenty milliliters of 0.2 M acetate buffers of pH 4, 3 and 2.2 were then passed through the column. Triiodothyronine was eluted with 50% acetic acid and the eluate collected in ten 3 ml fractions.[41] These were counted in a Packard Model 5213 Autogamma counter and those containing radioactivity were combined and extracted once with 15 ml of hexane in a 125 ml separatory funnel. The aqueous layer was transferred to a 100 ml beaker containing 3 mg potassium iodide to prevent deiodination[42] and was evaporated *in vacuo.*

Acetylation and Purification

The dry residue was rinsed three times with 0.3 ml of absolute ethanol and the washes transferred with a disposable pipette to a 12 ml dry glass-stoppered conical centrifuge tube. The ethanol was evaporated under nitrogen and the residue dried *in vacuo* over P_2O_5 for one hour. Absolute ethanol (0.25 ml) and 2 mg anhydrous sodium acetate was then placed in the tube. Ten microliters (0.5 mμM) of tritium-labeled acetic anhydride solution was added and the mixture incubated overnight at 23° C. The ethanol was then evaporated under nitrogen in the hood at 40° C and the residue dissolved in 2 ml of 4N sodium hydroxide containing 0.6 mg of stable N acetyl T3. Two milliliters of 10N sodium hydroxide was added dropwise and the mixture allowed to stand for ninety minutes during which the N-acetyl-T3 precipitated as the sodium salt. After centrifugation at 4000 rpm for ten minutes the clear supernatant was removed by aspiration. The precipitate was washed successively with 2 ml of 5 percent hydrochloric acid (v/v) and 2 ml of water.

Chromatography

The N-acetyl-T3 was then dissolved in 2 drops of ethanol/2N ammonium hydroxide (1:1) containing 5 mg 6-propyl-2-thiouracil/ml as an antioxidant [42] and spotted on a 20 x 20 cm glass plate coated with a 0.5 mm layer of Silica gel H (Brinkman Instruments). Two dimensional chromatography was done in tanks shielded from direct light with aluminum foil using freshly prepared solvent solutions: (1) butanol/ethanol/1N ammonium hydroxide, 100:20:30 v/v/v (BEA) and (2) butanol/water glacial acetic acid, 170:150:20 v/v/v, upper phase (BAc). After the first chromatography the plate was placed in the second tank as soon as drying was complete to avoid air oxidation of the compound.

The N-acetyl-T3 spot, 1.5 cm in diameter, was visualized briefly under short-wave ultraviolet light, demarcated with a needle point, and aspirated into the upper end of a Pasteur pipette containing a fine glass wool plug previously washed with methanol. The compound was eluted into a conical 12 ml centrifuge tube with 1 ml of methanol. The methanol was evaporated under nitrogen at 40° C and the N-acetyl-T3 promptly spotted on a second thin layer plate. Two dimensional chromatography was done using chloroform/methanol/28 percent ammonia, 100:50:5 (v/v/v), by means of two runs in the first and one run in the second direction. Slurries for these plates were prepared using methanol containing 0.001 M $Na_2S_2O_3$.[17] The spot was visualized and aspirated as before, followed by elution with 1 ml methanol into a glass counting vial.

Radioisotope Counting and Calculation of T3 Value

Ten milliliters of scintillation fluid[37] was added and each sample was counted for ten minutes in a dual channel Packard Model 3375 liquid scintillation spectrometer to a level of less than 1% counting error for ^{3}H and less than 2% for ^{131}I. Triiodothyronine content in micrograms per 100 ml was calculated by means of the following equations:

$$\frac{\dfrac{\text{cpm }^{3}\text{H} - (\text{cpm }^{3}\text{H bcg} + \text{cpm }^{131}\text{I crossover}) \times 100}{\%\text{ efficiency for }^{3}\text{H}}}{\%\ ^{131}\text{I recovery}/100} = \text{dpm Ac-T3-}^{3}\text{H} \quad (1)$$

$$\frac{\text{dpm Ac-T3-}^{3}\text{H} \times 20}{\text{sp. act. Ac-T3-}^{3}\text{H}} \times 0.94 = \mu\text{g triiodothyronine/100 ml serum} \quad (2)$$

The ^{3}H background (bcg) was obtained by eluting and counting a blank spot from a thin-layer plate run simultaneously with the assay plates in the final chromatography. A ^{3}H-toluene internal standard was used to determine counting efficiency. ^{131}I-crossover refers to the amount of ^{131}I tracer activity appearing in the ^{3}H channel. With appropriate settings of the liquid scintillation spectrometer this represented about 10 percent of the tracer counts, as indicated by counting crossover of the ^{131}I standard. ^{3}H counts were excluded from the ^{131}I channel. The ^{131}I recovery was calculated as a percent of the mean value of the ^{131}I-T3 standard duplicates. The specific activity of the tritiated T3 derivative was expressed as disintegrations per minute per microgram. The factor 0.94 corrects for the difference between the molecular weights of T3 and its N-acetyl derivative. The result was customarily expressed as nanograms of T3 per 100 ml of serum.

Determination of Specific Activity of N-acetyl T3-³H

The specific activity of the T3 derivative was verified experimentally (Fig. 7–1). Tritiated acetic anhydride (0.5 ml) was used to acetylate 2 mg of T4 in 1 ml of ethanol as previously described using proportionally reduced amounts of solutions and reagents.[37] The N-acetyl-T4 was recrystallized from 0.25 ml ethanol with 0.8 ml of water. The specific activity of N-acetyl-T4

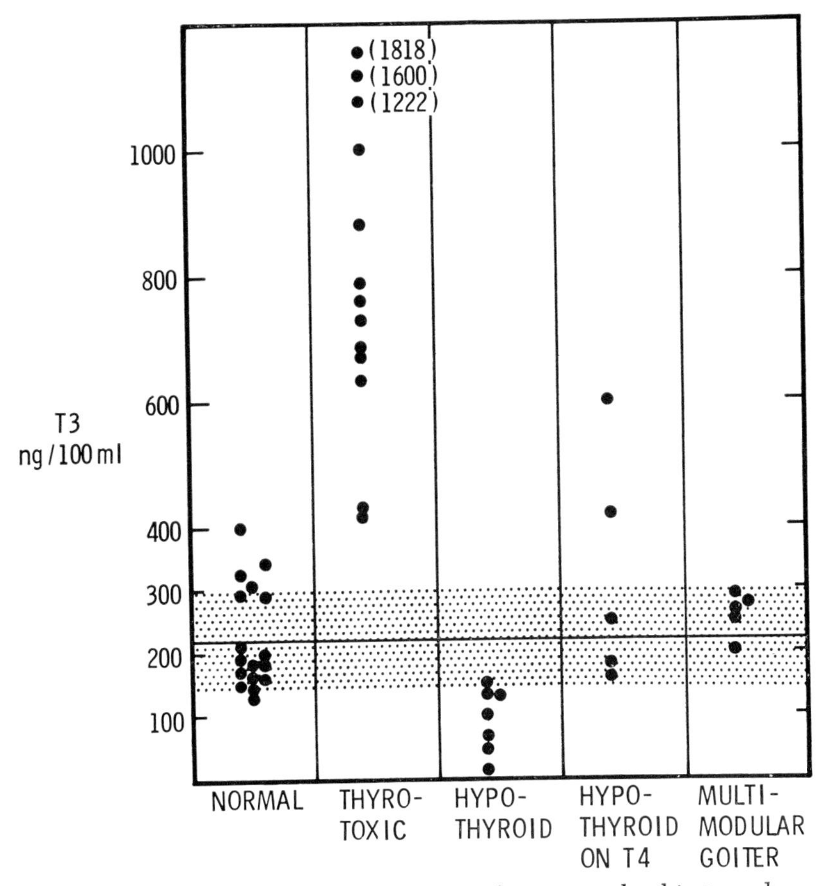

Figure 7–2. Results of T3 assays of serum from normal subjects and patients with thyroid disease. (Reproduced with permission from G.A. Hagen, L.I. Diuguid, B. Kliman, and J.B. Stanbury: Double isotope derivative assay of iodothyronines. III. Triiodothyronine. Biochem. Med. 7.191, 1973. *Biochem Med.*)

must be multiplied by 1.18 to obtain the value for the T3 derivative.

RESULTS

Results of assays of serum from normal subjects and patients with abnormal thyroid function are shown in Figure 7–2. Specificity and accuracy were demonstrated by assay of increments of T3 added to serum: 40, 200, 400, 600 and 800 ng/100 ml (Fig. 7–3). Precision of replicate assays is ± 14 percent. The results are unaffected by T4 either by deiodination to form T3 or

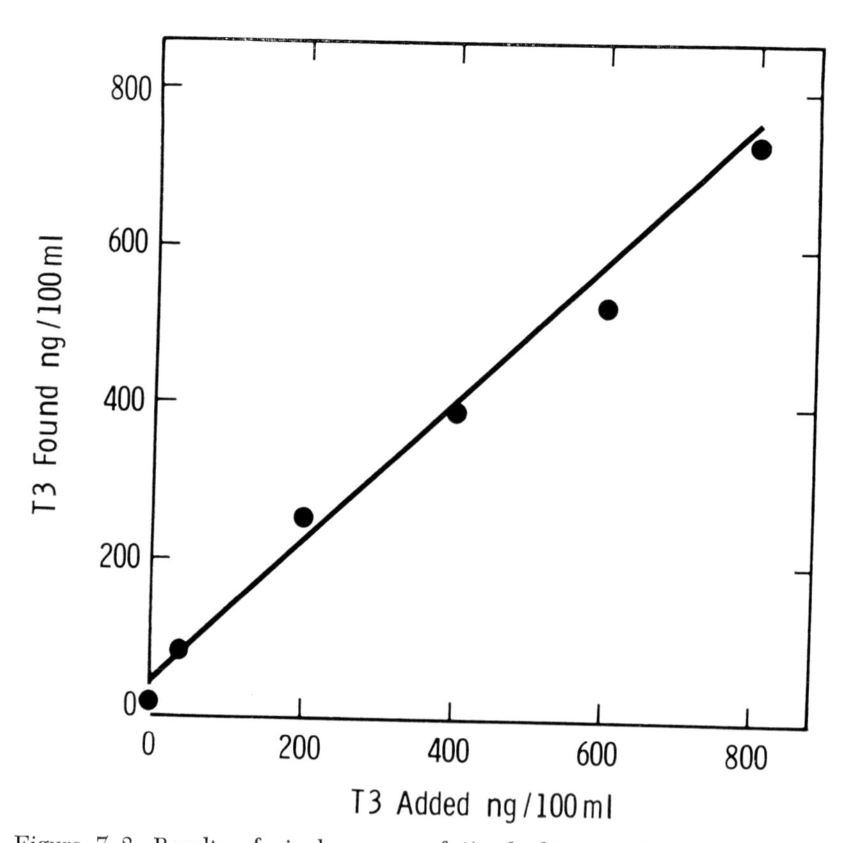

Figure 7–3. Results of single assays of 5 ml aliquots of serum from a hypothyroid patient containing increments of T3 added *in vitro.* (Reproduced with permission from G.A. Hagen, L.I. Diuguid, B. Kliman, and J.B. Stanbury: Double isotope derivative assay of iodothyronines. III. Triiodothyronine. Biochem. Med. 7:191, 1973. *Biochem Med.*)

incomplete separation from T3. Repeated trials of the assay substituting T4 ^{131}I for T3 ^{131}I as the tracer substance showed no recovery of ^{131}I after the final chromatography, thus obviating consideration of error due to intraassay formation of T3 from T4.

DISCUSSION

Methodologic Considerations in Double Isotope Derivative T3 Assay

The major problems encountered in the development of a double isotope derivative assay for T3 include (1) extraction of T3 from serum and readying it for acetylation, (2) elucidation of optimum conditions for the acetylation reaction, (3) preparation and purification of an acetylated derivative of known structure, (4) integration in the assay of a high specific activity T3 tracer labeled with a second isotope.

Verification of Derivatives

In order to lay a sound foundation for the T4 and T3 assays, methods of preparation and purification of the acetyl derivatives were established with nonradioactive compounds. Techniques were developed for purification of T3 and T4 obtained from the manufacturer. Mono- and diacetyl derivatives of T3 and T4 were prepared and their identity established in comparison with authentic parent compounds by means of Rf values, acid-base titration curves, nitrous acid reaction, melting points, elemental analysis and infrared spectroscopy (Fig. 7–4).[36] The possibility of error due to misinterpretation of the nature of the assay derivative was thereby eliminated. Study of the acetyl derivatives of T3 and T4 revealed the monoacetyl to be most appropriate for use in the assay since it was more stable in acid and alkaline solutions than the diacetyl, and also easier to purify by chromatography because of its intermediate Rf value (Fig. 7–4).

Adaptation of T4 Method

Our next approach involved development *de novo* of an assay for T4 which circulates at a higher level and therefore is pre-

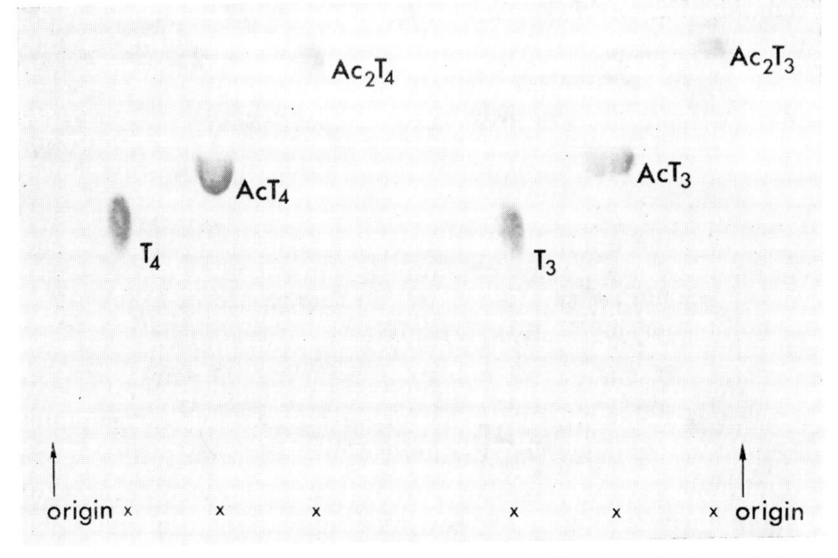

Figure 7–4. Thin layer chromatogram of purified T3 and T4 and their mono and diacetyl derivatives. A 20 x 20 cm glass plate coated with 0.5 mm of Silica gel H was developed in butanol:ethanol:1N ammonium hydroxide, 100:20:30 v/v/v.

sumably easier to measure. Extraction of T4 from serum with an anion exchange resin column was done prior to acetylation. Although pyridine has been used for many years as an acetylation solvent, we have found absolute ethanol distilled after refluxing with sodium methoxide and diethyl phthalate [37] to be more satisfactory for either T3 or T4 at 23° C. Benzene and similar solvents cannot be used because of solubility problems. After acetylation of the unknown T4, stable N-acetyl-T4 is added to serve as a carrier and marker. A few simple precipitation steps are employed and thin layer chromatography is then used for the balance of the purification. Once the T4 method was established and its accuracy and specificity verified (Fig. 7–5), it was modified for measurement of T3.

Use of Papain Digestion

Solution of the remaining problems can be discussed in terms of the completed T3 assay method as outlined in Figure 7–6.

$T3-^{131}I$ obtained from Abbott Laboratories was purified of contaminating iodide by Sephadex column adsorption and stored as previously described.[37] Chromatography and radioautography of stored $T3-^{131}I$ has shown no evidence of deterioration for at

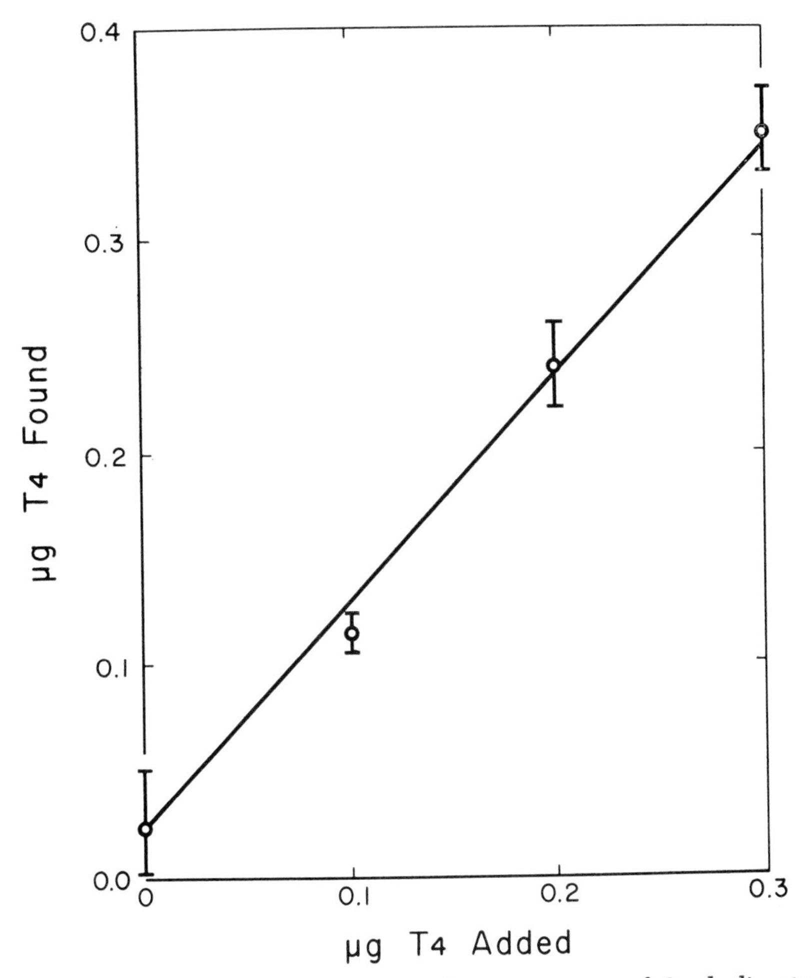

Figure 7–5. Results of double isotope derivative assays of 2 ml aliquots of serum from a hypothyroid patient containing increments of T4 added *in vitro*. Each value represents the mean of 4 assays ± S.D. (Reproduced with permission from G.A. Hagen, L.I. Diuguid, B. Kliman, and J.B. Stanbury, Double isotope derivative assay of iodothyronines. II. Thyroxine. *Anal Biochem*, 38:517, 1970.)

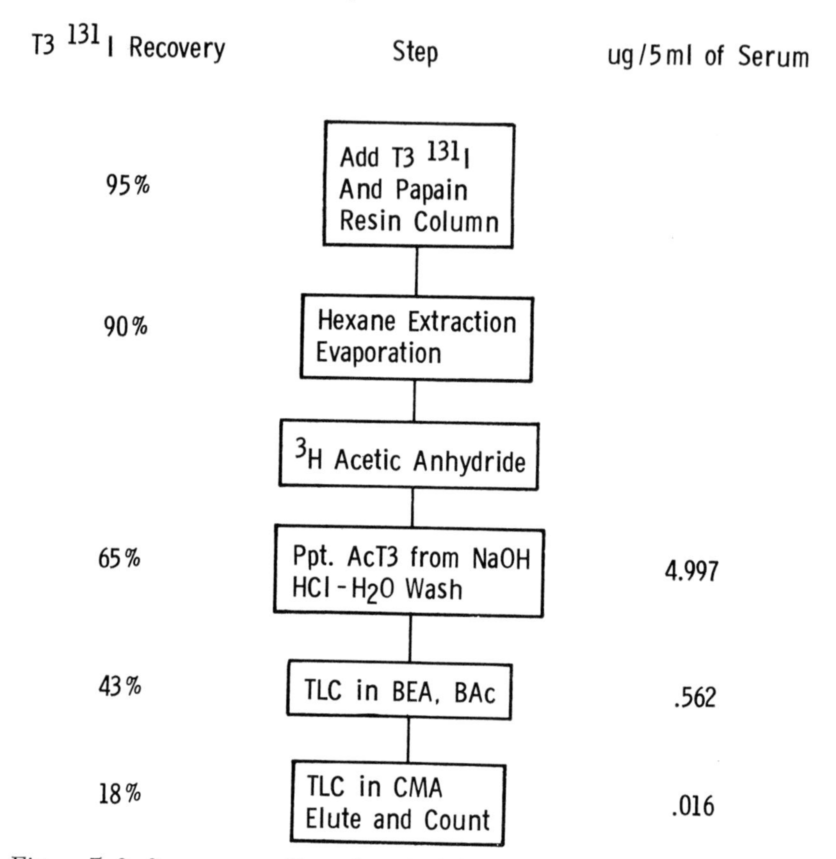

Figure 7–6. Summary outline of method for T3 assay illustrating stepwise purification of acetyl derivative and corresponding tracer recovery. (Reproduced with permission from G.A. Hagen, L.T. Diuguid, B. Kliman, and J.B. Stanbury: Double isotope derivative assay of iodothyronines. III. Triiodothyronine. Biochem. Med. 7.191, 1973. *Biochem Med.*)

least six weeks. The specific activity is such that 10,000 cpm added to 5 ml of serum constitutes less than 6 ng T3 per 100 ml. Following incubation with the tracer, the serum sample is digested with papain.[40] Enzymatic digestion prior to extraction was found to improve final recovery and decrease the number of purification steps subsequent to acetylation. Chromatography of eluates of T3–^{131}I from the resin column after serum extraction revealed a variable small fraction of the labeled hormone to be present as a nondissociable complex similar to that recently reported by Oppenheimer et al.[43] Papain digestion prior to extraction appeared to prevent the occurrence of this complex with consequent improved T3 recovery.

High Specific Activity Tritium-labeled Acetic Anhydride

In the adaptation of the T4 double isotope derivative assay for T3, two new major problems were encountered: (1) use of tritium-labeled acetic anhydride with a specific activity in the range of 1 to 3 Ci/mM and (2) separation of T3 from T4. The fact that T3 circulates at a level of about one thirtieth that of T4 makes it much more difficult to assay. An adequately sensitive assay is made possible by the availability of tritium-labeled acetic anhydride with a specific activity of 1 to 3 curies per millimole. Using this reagent an acetyl derivative of T3 can be prepared with a specific activity of 1 to 3 million dpm per microgram (Fig. 7–1). This plus a serum sample of 5 ml provides sufficient sensitivity for assay of serum T3. However, use of this reagent entails greater difficulties in purification due to its higher specific activity. Furthermore, the reagent itself must be of the highest chemical purity and carefully protected from adverse conditions. This is because the actual quantitative amount of acetic anyhdride used is much lower because of the greatly increased specific activity.

For example, tritium-labeled acetic anhydride with a relatively low specific activity of 100 millicuries per millimole can be used for T4 assay. An amount of acetic anhydride of about 0.02 millimole is used for acetylation. However, the highest specific activity acetic anhydride usually available has about 2000 millicuries per millimole. In this instance, only .5 millimicromole of tritium-labeled acetic anhydride was added to the acetylating solution, a much smaller amount. This was found to acetylate sufficient T3 in spite of other competing substances necessarily present in the acetylating solution. Use of larger quantities of acetic anhydride overloaded the purification procedure with tritium contamination of the final result. Improperly prepared or deteriorated acetic anhydride reagent was found to be the most frequent reason for low recovery or contamination of the assay results. Customarily, 10 microliters of a 0.5 percent solution of acetic anhydride in benzene was added to the solution in the acetylation tube. This amount of acetic anhydride acetylated about 60 percent of the T3 present.

Separation of T3 from T4

Secondly, and equally important, is the problem of achieving complete separation of T3 from T4 during the assay procedure. These compounds are difficult to separate completely chromatographically. Even 0.5 percent of T4 would be sufficient to

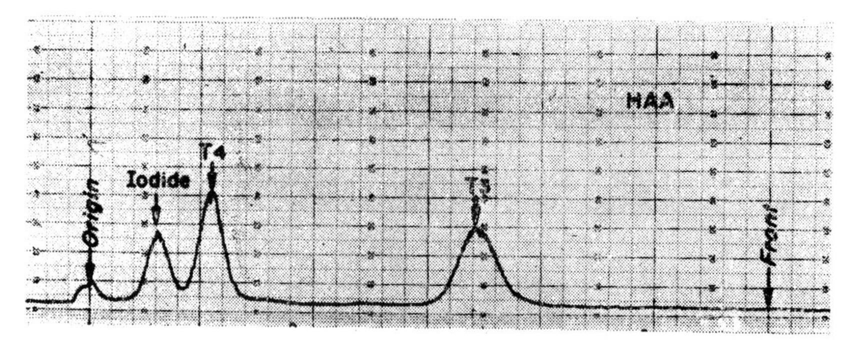

Figure 7–7. Descending paper chromatogram showing separation of T3 and T4 during T3 assay by the Sterling method. (Reproduced with permission from H. Wahner, and C. A. Gorman, Interpretation of serum triiodothyronine levels measured by the Sterling Technique. *N Engl J Med, 284*:225, 1971.)

invalidate the T3 results, so that complete separation must be achieved during the assay (Fig. 7–7). Use of chloroform/methanol/ammonia (CMA), a solvent system described by Heider and Bronk,[45] for a second thin layer chromatography, led to complete separation of the monoacetyl derivatives of T3 and T4 (Fig. 7–8). Thus if the T3 assay were run using a T4 ^{131}I tracer instead of T3 ^{131}I there was zero recovery of the T4 tracer in the T3 assay in several trials. This gave assurance that T3 and T4 were completely separated and that also there was no contamination of the T3 results from deiodination of T4 within the assay to form spurious amounts of T3.

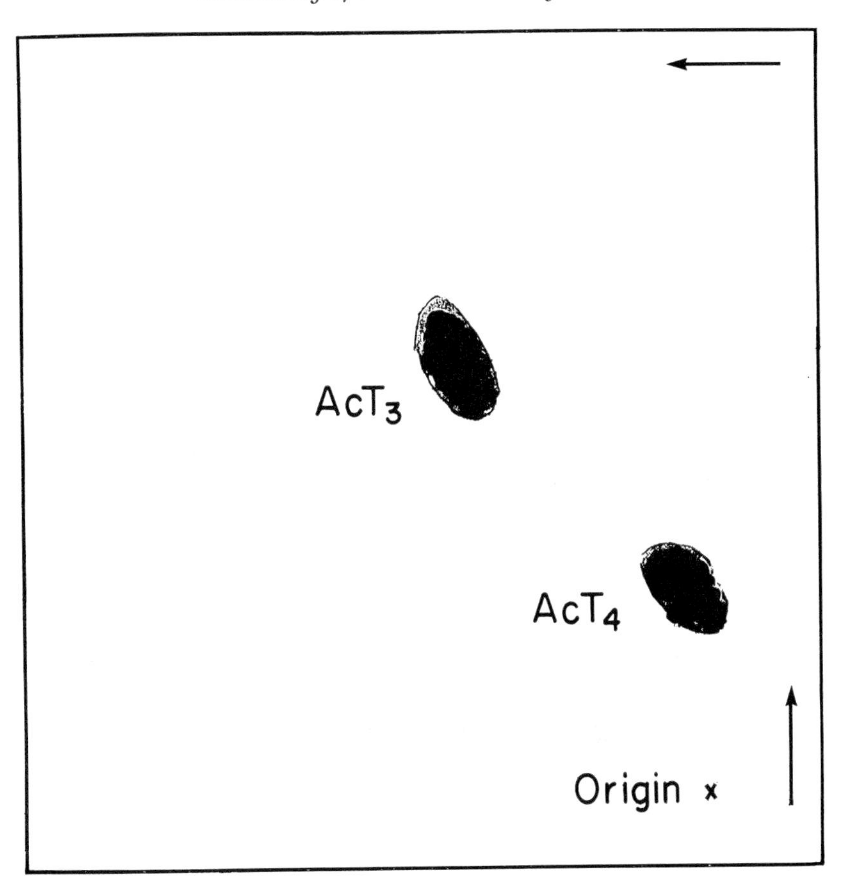

Figure 7–8. Two-dimensional thin layer chromatogram showing separation of the monoacetyl derivatives of T3 and T4 during a simultaneous T3–T4 assay by the double isotope derivative method. Silica gel H containing the derivatives was visualized and aspirated from the plate as described in the text. The solvent was chloroform/methanol/28% ammonia 100:-50:5 v/v/v.

Elimination of Intraassay Deiodination Artefact

There are a number of reasons for absence of contamination of the final T3 result from T4 either by incomplete separation or deiodination. AcT4 is only slightly soluble in 4N NaOH so that about 90 percent is excluded at this point in the assay. No significant separation of AcT3 from AcT4 is accomplished in the

BEA-BAc chromatography but unacetylated T4 is removed. The protective steps taken in CMA chromatography greatly decrease deiodination: (1) spotting with solvent containing PTU, (2) addition of $Na_2S_2O_3$ to the methanol while making the TLC slurry. Finally, wide separation of the n-acetyl T3 derivatives is accomplished in the CMA thin-layer chromatography; it is necessary to monitor the conditions of time, temperature and humidity to assure consistent chromatographic results. A comparison of the purification of the derivative with the recovery of the ^{131}I tracer is shown in Figure 7–6. Accuracy and specificity of the assay was proved by adding increments of purified T3 to serum from a myxedematous patient and recovering the increments as shown in the Figure 7–3. Such demonstration of recovery is greatly supportive in assessing the validity of the assay results. In addition, sera from myxedematous patients has been assayed to a level as low as 20 ng of T3/100 ml.

Comparison with Other Methods

Finally it can be seen (Table 7–I) that results of the double isotope derivative assay compare favorably in amount with those found by competitive protein binding techniques and also by some of the most recently reported radioimmunoassay methods. The importance of the double isotope derivative method lies in

TABLE 7–I
SERUM T3 VALUES REPORTED FOR VARIOUS METHODS OF QUANTITATION

	Serum T3, ng/100 ml		
	Normal	*Hypo*	*Hyper*
Pind *	256	—	—
Nauman et al.†	330	100	710
Hollander ‡	450	200	1210
Sterling et al.†	220	98	752
Hagen et al.§	226	92	897
Dussault et al.†	98	0	443
Ekins ‖	120	—	—
Gharib et al.‖	218	103	760
Chopra et al.‖	<100	<100	519
Mitsuma et al.‖	138	62	494
Lieblich et al.‖	145	99	429

* Iodine by ceric sulfate-arsenious acid reaction
† Competitive protein binding
‡ Gas chromatography
§ Double isotope derivative
‖ Radioimmunoassay

the fact that it is a chemically specific method by which the substance to be assayed is directly identified and measured. This is to say, one does not depend upon a dose response curve involving protein binding. Although this method may not be as convenient as a successful radioimmunoassay, we feel that it will be of importance because of its reliability and consistency. The method as we have developed it does compare well with the length of the procedures used for competitive protein binding. It is not necessary to vary the amount of serum at low levels of T3; also, no correction for methodological conversion of T4 to T3 is required.

Radioimmunoassay of Serum T3

Until recent years radioimmunoassay was not attempted for nonpeptide hormones such as T3 or T4 since such small molecules were not immunogenic. However, workers have found that with compounds such as angiotensin or digitoxin, antibodies could be prepared in the following manner: The compound to be assayed is first coupled to another antigenic substance such as poly-L-lysine or albumin.[46,47] Injection of animals with such a substance may eventually result in formation of an antibody that will react with the unknown compound alone with sufficient sensitivity for assay purposes.

In 1970 Ekins and coworkers reported successful formation of antibodies to triiodothyronine by linking it to poly-L-lysine.[48] Gharib et al. were unable to obtain antibodies by this method but were successful after linking T3 to albumin by a diimide method (Fig. 7–9).[49] Chopra et al. obtained antibodies to T3 with the use of thyroglobulin as an antigen.[50] If an effective antiserum can be obtained and a dose response curve set up, performance of radioimmunoassay of T3 or T4 can be anticipated. However, some of the dose response curves published are sensitive over a rather narrow bound/free range, for example, between 40 and 50 percent in the assay reported by Mitsuma et al.[51]

Secondly, antibody reactivity must be preserved from competition with the binding proteins already present in serum, especially thyroxine binding globulin (TBG). In some instances this problem has also been solved by saturating the T3 binding

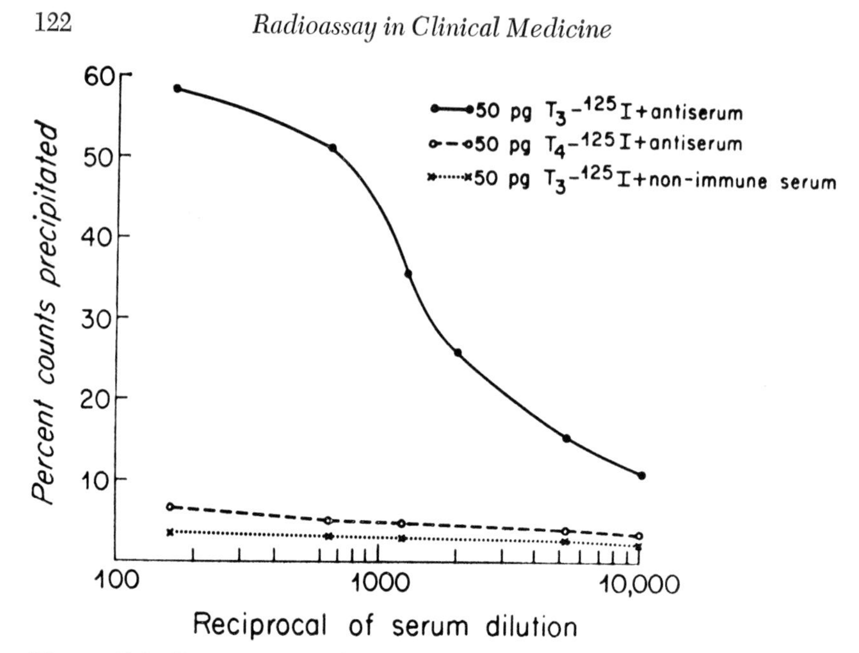

Figure 7–9. Precipitation of antibody-bound T3^{125}I. Progressively decreased dilution of antiserum. No binding of labeled T4 is seen; non-immune serum does not react. (Reproduced with permission from H. Gharib, W.E. Mayberry, and R.J. Ryan, *J Clin Endocrinol,* Radioimmunoassay for triiodothyronine: A preliminary report. *31*:709, 1970.)

serum globulin with another substance such as T4 itself, tetrachlorothyronine, or diphenylhydantoin. In general, it has not been conclusively shown that all T3 is displaced from serum-binding proteins for reaction with the antibody. Incomplete displacement would result in falsely low values. Preliminary results have been reported from a number of laboratories (Table 7–I). In general, results of T3 immunoassay have been lower than the levels reported by earlier methods. Indeed, one method is unable to measure T3 except at thyrotoxic levels.[50] Another radioimmunoassay method finds normal serum T3 values to be similar to those obtained by the Sterling method.[52] The reason for the discrepancy between this and other immunoassays is not apparent. Three laboratories report mean normal T3 serum levels by radioimmunoassay in the range of 120 to 145 ng/100 ml.[51,53,54] Thus a conclusive assessment as to the accuracy and utility of T3 immunoassay must await the results of further experience with the method. Clinical investigations employing T3 radioimmunoassay have been initiated; in four patients who developed thyrotoxicosis serum T3 was found to rise prior to T4.[55] Elevation of human serum T3 and TSH in response to intravenous thyrotropin releasing factor has been observed.[56]

Clinical Application of T3 Assays—"T3 Toxicosis"

Serum T3 assay is of primary interest to the clinician in connection with patients who may have "T3 toxicosis." Such patients may have diffuse or nodular goiters or autonomous nodules.[25,31,32,57,58] Characteristically they may present with some clinical features of thyrotoxicosis yet are found to have normal serum thyroxine levels. At times there is a prolonged history of recurrent cardiac atrial arrhythmia. The following summary is an example of a patient with T3 toxicosis.

> A 40-year-old woman was admitted to the St. Louis University Hospital with a six-month history of an enlarging goiter, weakness, nervousness, tremors and palpitation. There had been one recent episode of probable paroxysmal tachycardia. Physical examination revealed a multinodular goiter which was three times normal size. The patient appeared to be clinically euthyroid. An electrocardiogram revealed multifocal premature ventricular contractions. PBI was 5.0 μg/100 ml (n = 3.5 to 7.5) and the serum T4 by the Murphy-Pattee method was 5.4 (n = 5.0 to 13.7). The T4 resin binding index was 0.99 (n = 1.13 to 0.87). The 6- and 24-hour thyroidal radioiodine uptake was 9% (n = 7 to 25) and 15% (n = 15 to 40). Serum T3 by the double isotope derivative method was 1300 ng/100 ml (n = 100 to 300); a second sample contained 1000 ng/100 ml.
>
> Because of the history of probable cardiac arrhythmia, a T3-suppression text was not done. A diagnosis of T3 thyrotoxicosis was made and subtotal thyroidectomy performed. Postoperative follow-up revealed no recurrence of the arrhythmia or symptoms described.

If one is presented with such a patient there are a number of clinical tests that can go a long way toward establishing a diagnosis before an actual quantitative T3 serum level is sought. They are as follows (Table 7–II):

TABLE 7–II
LABORATORY STUDIES OF IMPORTANCE FOR THE DIAGNOSIS OF T3 TOXICOSIS

1. RAI, T4, *BMR*
2. *T3 suppression test*
3. free T4, *TBG*, TBPA
4. *Total T3*

Basal Metabolic Rate

This test when properly done is of great assistance in that it supports the clinician's impression of hypermetabolism.[59] It points the way towards further investigation.

T3 Suppression Test

Normally the administration of 75 to 100 micrograms of T3 over a period of seven days will result in at least a 50% reduction of the twenty-four-hour radioiodine uptake. If the patient already has a high circulating level of T3 the RAI uptake will not be suppressed by giving more T3. Demonstration of nonsuppression adds further support to the diagnosis of T3 toxicosis.

It must also be noted that there are some patients who have decreased T4-binding capacity of circulating thyroxine-binding globulin which may result in a depressed T4 in the presence of thyrotoxicosis.[60] Thus, if the first two measures are positive, a serum TBG level should be obtained. Finally, demonstration of circulating T3 level from the quantitative standpoint is of great usefulness in making the final diagnosis. There has been some recently presented evidence that at times in patients who are followed long enough the serum T4 may come to be elevated as well as serum T3.[55] Treatment of such patients follows the same general principles already in effect for the more common types of thyrotoxicosis.

CONCLUSION

No doubt elucidation of the action of T3 in thyroid metabolism will alter many current concepts of thyroid pathophysiology. Although it appears to be of at least equal significance to T4 in body metabolism, the role of T3 in terms of its formation, distribution, utilization and breakdown requires further clarification. The double isotope derivative method described can be adapted for simultaneous assay of T3 and T4.[61] Such an approach may provide information concerning the relationship of these two hormones in various diseases of the thyroid gland and after therapy of thyrotoxicosis, especially with radioactive iodine therapy. It may also be pertinent to estimation of the ability of the thyroid to adapt to conditions of iodine deficiency.

Conceivably, illumination of the role of T3 by the new assays may be of importance in certain cardiovascular states, especially the refractory supraventricular arrhythmias. Replacement therapy in thyroid-deficient states may be improved and unwanted side effects eliminated when a more physiologic dose of T3 or T4 is determined. Assay of T3 and T4 may be used to develop additional information concerning the relative rates of secretion of these hormones by the thyroid gland. The ultimate role of the individual T3 assays remains to be determined; however it would appear that radioimmunoassay will prove to be more convenient than the double isotope derivative technique which may serve as a reference method.

REFERENCES

1. Kendall, E.C.: Isolation of the iodine compound which occurs in the thyroid. *J. Biol. Chem. 39*:125, 1919.
2. Gross, J. and Pitt-Rivers, R.: Identification of 3:5:3′-L-triiodothyronine in human plasma. *Lancet 1*:439, 1952.
3. Gross, J.; Pitt-Rivers, R. and Trotter, W.R.: Effect of 3:5:3′-L-triiodothyronine in myxoedema. *Lancet 1*:439, 1952.
4. Lerman, J.: Physiologic activity of L-triiodothyronine. *J. Clin. Endocr. 13*:1341, 1953.
5. Pitt-Rivers, R.; Stanbury, J.B. and Rapp, B.: Conversion of thyroxine to 3-5-3′-triiodothyronine in vivo. *J. Clin. Endocr. 15*:616, 1955.
6. Lassiter, W.E. and Stanbury, J.B.: The in vivo conversion of thyroxine to 3:5:3′-triiodothyronine. *J. Clin. Endocr. 18*:903, 1958.
7. Benua, R.S. and Dobyns, B.M.: Iodinated compounds in serum, disappearance of radioactive iodine from thyroid, and clinical response in patients treated with radioactive iodine. *J. Clin. Endocr. 15*:118, 1955.
8. Dingledine, W.S.; Pitt-Rivers, R. and Stanbury, J.B.: Nature and

transport of iodinated substances of blood of normal subjects and of patients with thyroid disease. *J. Clin. Endocr. 15*:724, 1955.

9. Benua, R.S.; Dobyns, B.M. and Ninmer, A.: Triiodothyronine in serum of patients treated with radioactive iodine. *J. Clin. Endocr. 15*:1367, 1955.
10. Maclagan, N.F.; Dowden, C.H. and Wilkinson, J.H.: The metabolism of thyroid hormones. 2. Detection of thyroxine and triiodothyronine in human plasma. *Biochem. J. 67*:5, 1957.
11. Arons, W.L. and Hydovitz, J.D.: The serum pattern of thyroid hormones in euthyroidism and hyperthyroidism. *J. Clin. Endocr. 19*:548, 1959.
12. Rupp, J.J.; Chavarria, C.; Paschkis, K.E. and Chublarian, E.: The occurrence of tri-iodothyronine as the only circulating thyroid hormone. *Ann. Int. Med. 51*:359, 1959.
13. Werner, S.C.; Row, V.V. and Radichevich, I.: Nontoxic nodular goiter with formation and release of a compound with the chromatographic mobility characteristics of triiodothyronine. *J. Clin. Endocr. 20*:1373, 1960.
14. Mack, R.E.; Hart, K.T.; Druet, D. and Bauer, M.A.: An abnormality of thyroid hormone secretion. *Am. J. Med. 30*:323, 1961.
15. Rupp, J.J. and Paschkis, K.E.: The changing pattern of circulating iodinated amino acids in a case of thyrotoxicosis. *Am. J. Med. 30*:472, 1961.
16. Shimaoka, K.: Toxic adenoma of the thyroid, with triiodothyronine as the principal circulating thyroid hormone. *Acta Endo. 43*:285, 1963.
17. Pind, K.: Paperchromatographic determination of thyroid hormone (3, 5, 3'- triiodothyronine) in serum without radio-iodine. *Acta Endo.*(Kbh), *26*:263, 1957.
18. Shimaoka, K. and Jasani, B.M.: The application of two-dimensional paper chromatography and low-level counting to the study of triiodothyronine in plasma. *J. Endocr. 32*:59, 1965.
19. Rawson, R.W.; Rall, J.E.; Pearson, O.H.; Robbins, J.; Poppell, H.F. and West, C.D.: 1-triiodothyronine versus 1-thyroxine: comparison of their metabolic effects in human myxedema. *Am. J. Med. Sci. 226*:405, 1953.
20. Robbins, J. and Rall, R.E.: Proteins associated with the thyroid hormones. *Physiol. Rev. 40*:415, 1960.
21. Nauman, J.A.; Nauman, A. and Werner, S.C.: Total and free triiodothyronine in human serum. *J. Clin. Invest. 46*:1346, 1967.
22. Sterling, K.; Bellabarba, D.; Newman, E.S. and Brenner, M.A.: Determination of triiodothyronine concentration in human serum. *J. Clin. Invest. 48*:1150, 1969.
23. Larsen, P.R.: Technical aspects of the estimation of triiodothyronine in human serum: evidence of conversion of thyroxine to triiodothyronine during assay. *Metabolism 20*:609, 1971.

24. Dussault, J.H.; Lam, R. and Fisher, D.A.: The measurement of serum triiodothyronine by double column chromatography. *J. Lab. Clin. Med. 77:*1039, 1971.
25. Hollander, C.S.: On the nature of the circulating thyroid hormone. Clinical studies of triiodothyronine and thyroxine in serum using gas chromatographic methods. *Trans. Assoc. Am. Phys. 81:*76, 1968.
26. Braverman, L.E.; Ingbar, S.H. and Sterling, K.: Conversion of thyroxine (T4) to triiodothyronine (T3) in athyreotic human subjects. *J. Clin. Invest. 49:*855, 1970.
27. Sterling, K.; Brenner, M.A. and Newman, E.S.: Conversion of thyroxine to triiodothyronine in normal human subjects. *Science 169:*1099, 1970.
28. Pittman, C.S.; Chambers, J.B. and Read, V.H.: The extrathyroidal conversion rate of thyroxine to triiodothyronine in normal man. *J. Clin. Invest. 50:*1187, 1971.
29. Sterling, K.: The significance of circulating triiodothyronine. *Recent Prog. Horm. Res. 26:*249, 1970.
30. Woebar, K.A.; Sobel, R.J.; Ingbar, S.H. and Sterling, K.: The peripheral metabolism in triiodothyronine in normal subjects and patients with hyperthyroidism. *J. Clin. Invest. 49:*643, 1970.
31. Sterling, K.; Refetoff, S. and Selenkow, H.A.: T3 thyrotoxicosis. Thyrotoxicosis due to elevated serum triiodothyronine levels. *J. Am. Med. Assoc. 213:*571, 1970.
32. Ivy, H.K.; Wahner, H.W. and Gorman, C.A.: Triiodothyronine (T3) toxicosis: Its role in Graves' disease. *Arch. Int. Med. 128:*529, 1971.
33. Udenfriend, S.: Identification of α-aminobutyric acid in brain by isotope derivative method. *J. Biol. Chem. 187:*65, 1950.
34. Kliman, B. and Peterson, R.E.: Double isotope derivative assay of aldosterone in biological extracts. *J. Biol. Chem. 235:*1639, 1960.
35. Whitehead, J.K. and Beale, D.: The determination of thyroxine levels in human plasma by double isotope-dilution technique. *Clin. Chim. Acta 4:*710, 1959.
36. Hagen, G.A.; Diuguid, L.I.; Kliman, B. and Stanbury, J.B.: Double-isotope derivative assay of serum iodothyronine. I. Preparation of acetyl derivatives of thyroxine and triiodothyronine. *Anal. Biochem. 33:*67, 1970.
37. Hagen, G.A.; Diuguid, L.I.; Kliman, B. and Stanbury, J.B.: Double-isotope derivative assay of serum iodothyronines. II. Thyroxine. *Anal. Biochem. 38:*517, 1970.
38. Hagen, G.A.; Diuguid, L.I.; Kliman, B. and Stanbury, J.B.: Double-isotope derivative assay of serum triiodothyronine. *Clin. Res. 18:*602, 1970.
39. Hagen, G.A.; Diuguid, L.I.; Kliman, B. and Stanbury, J.B.: Double-isotope derivative assay of serum iodothyronines. III. Triiodothyronine. *Biochem. Med. 7:*191, 1973.

40. Lissitzky, S.; and Bismuth, J.: Quantitative determination of the concentration of I-131-thyroxin and I-131-triiodothyronine in the serum by filtration on dextran gel (Sephadex). *Clin Chim. Acta 8:*269, 1962.
41. Galton, V.A. and Pitt-Rivers, R.: A quantitative method for the separation of thyroid hormones and related compounds from serum and tissues with an anion-exchange resin. *Biochem. J. 72:*310, 1959.
42. Reinwein, D. and Rall, J.E.: Nonenzymatic deiodination of thyroid hormones by flavin mononucleotide and light. *J. Biol. Chem. 241:*1636, 1966.
43. Morreale De Escobar, G.; Llorente, P.; Jolin, T. and Escobar Del Ray, F.: The 'transient instability' of thyroxine and its biochemical applications. *Biochem. J. 88:*526, 1963.
44. Oppenheimer, J.H.; Surks, M.I.; Kozyreff, V.; Riba, A. and Koerner, D.: In vitro formation of nondissociable thyroid hormone-protein complexes. *J. Clin. Invest. 50:*241a, 1971.
45. Heider, J.G. and Bronk, J.R.: A rapid separation of thyroxine and some of its analogues by thin-layer chromatography. *Biochem. Biophys. Acta 95:*353, 1965.
46. Haber, E.; Page L.B. and Jacoby, G.A.: Synthesis of antigenic branch-chain copolymers of angiotensin and poly-L-lysine. *Biochemistry 4:*693, 1965.
47. Oliver, G.C. Jr.; Parker, B.M.; Brasfield, D.L. and Parker, C.W.: The measurement of digitoxin in human serum by radioimmunoassay. *J. Clin. Invest. 47:*1035, 1968.
48. Brown, B.L.; Ekins, R.P.; Ellis, S.M. and Reith, W.S.: Specific antibodies to triiodothyronine hormone. *Nature 226:*359, 1970.
49. Gharib, H.; Mayberry, W.E. and Ryan, R.J.: Radioimmunoassay for triiodothyronine: A preliminary report. *J. Clin. Endocr. 31:*709, 1970.
50. Chopora, I.J.; Solomon, D.H. and Beall, G.H.: Radioimmunoassay for measurement of triiodothyronine in human serum. *J. Clin. Invest. 50:*2033, 1971.
51. Mitsuma, T.; Nihei, N.; Gershengorn, M.C. and Hollander, C.S.: Serum triiodothyronine: Measurements in human serum by radioimmunoassay with corroboration by gas-liquid chromatography. *J. Clin. Invest. 50:*2679, 1971.
52. Gharib, H.; Mayberry, W.E.; Hocker, T. and Ryan R.J.: Radioimmunoassay of triiodothyronine (T3)9, page A-36, Program of Endo. Society 53rd Meeting, San Francisco, 1971.
53. Ekins, R.P.; Brown, B.L.; Ellis, S.M. and Reith, W.S.: The radioimmunoassay of serum triiodothyronine. Proc. 6th Int. Conf., p. 138, 1970.
54. Lieblich, J.M. and Utiger, R.B.: Triiodothyronine radioimmunoassay. *J. Clin. Invest. 51:*157, 1972.
55. Hollander, C.S.; Shenkman, L.; Mitsuma, T.; Blum, M.; Kastin, A.J.

and Anderson, D.G.: Hypertriiodothyroninaemia as a premonitory manifestation of thyrotoxicosis. *Lancet 2*:731, 1971.

56. Shenkman, L.; Mitsuma, T.; Suphavai, A. and Hollander, C.S.: Triiodothyronine and thyroid-stimulating hormone respone to thyrotrophin-releasing hormone. *Lancet 1*:111, 1972.
57. Wahner, H. and Gorman, C.A.: Interpretation of serum tri-iodothyronine levels measured by the Sterling technique. *New Eng. J. Med. 284*:225, 1971.
58. Radichevich, I. and Werner, S.C.: Increased serum concentrations of triiodothyronine in euthyroid patients with "hot" thyroid nodules. Proc. Ann. Meeting Amer. Thyroid Assoc., p. 13, 1968.
59. Keating, F.R.: In defense of the basal metabolic level. *J. Clin. Endocr. 17*:797, 1957.
60. Braverman, L.E.; Foster, A.E. and Ingbar, S.H.: Thyroid hormone transport in the serum of patients with thyrotoxic Graves' disease before and after treatment. *J. Clin. Invest. 47*:1349, 1968.
61. Hagen, G.A.; Diuguid, L.I.; Kliman, B. and Stanbury, J.B.: A chemically specific assay for serum triiodothyronine. Program, Fourth International Congress of Endocrinology, Washington, D.C., June, 1972. Abstract No. 608.

CHAPTER 8

LONG-ACTING THYROID STIMULATOR BIOASSAY

Francis A. Zacharewicz

BACKGROUND

THE SPECIFIC RELATIONSHIP of clinical symptoms with disease of the thyroid has been recognized since 1870. Thyroid replacement therapy was demonstrated to relieve the clinical symptoms of myxedema as early as 1890. The concept that the anterior pituitary controlled the thyroid through a tropic hormone (thyrotropin, thyroid stimulating hormone, TSH) was not developed until 1930. Subsequently, the pituitary-thyroid axis and in particular, the negative feedback mechanism was described by Salter [1] in 1940. This mechanism, as known today, is illustrated in Figure 8–1. The recognized steps in this biosynthetic process are trapping, oxidation, binding, coupling and finally the release of hormone. Iodide ingested in the diet is trapped in the thyroid, whereupon the cells oxidize it from iodide to some higher valence form, possibly iodine (I_2). This step occurs rapidly and perhaps simultaneously with the organic binding of the amino acid tyrosine, producing monoiodotyrosine (MIT) and diiodotyrosine (DIT). The next step involves the coupling of MIT and DIT in some way to form thyroxine and triiodothyronine. The hormone is stored in the form of thyroglobulin inside the follicles and broken down by proteolytic enzymes before entry into the general circulation. This entire process is stimulated by thyrotropin (TSH). A decrease in the circulating level of thyroid hormone stimulates the hypothalamic-pituitary axis to release TSH which then acts upon the thyroid gland. The hypothalamus releases a hormone-thyrotropin releasing factor (TRF) which stimulates the anterior pituitary to secrete TSH.[2] Conversely, an increase in

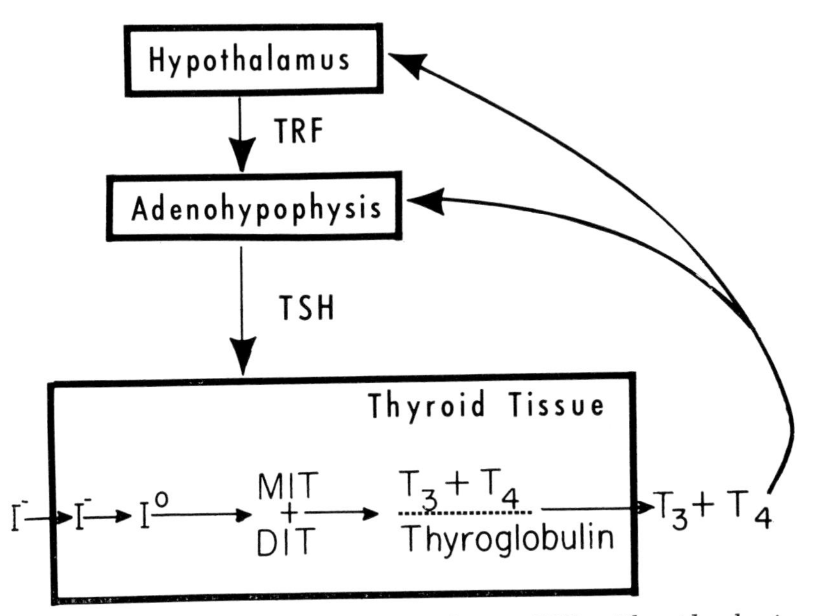

Figure 8–1. Schema of pituitary thyroid axis. TRF = Thyroid releasing factor (hormone); TSH = Thyroid stimulating hormone; MIT = Moniodotyronine; T_4 = Thyroxine. See text for explanation.

circulating thyroid hormone diminishes the activity of the hypothalamic-pituitary axis and TSH secretion falls. This feedback mechanism maintains the patient in an euthyroid state.

With this basic information, it was natural to assume that the mechanism responsible for producing thyrotoxic Graves' disease was uncontrolled or excessive release of thyrotropin. Various bioassay techniques [3] were developed, but failed to consistently demonstrate elevations of TSH levels in patients with Graves' disease (Table 8-I). This failure was interpreted to be due to the insensitive and nonspecific nature of the assay techniques. With the advent of the immunoassay [4,5] for the determination of circulating TSH, the initial findings of the bioassay measure-

TABLE 8–I
BIOASSAY VALUES OF TSH IN MAN

Normal—0.2 mU/ml plasma
Hyperthyroidism—widely scattered values
Hypothyroidism—up to 4 mU/ml plasma

TABLE 8-II
RADIOIMMUNOASSAY VALUES OF TSH IN MAN

Normal—< 3 mμg/ml
Hyperthyroidism—no difference from normal
Hypothyroidism—10 to 160 mμg/ml

ments were confirmed (Table 8-II). TSH levels were consistently elevated above normal in patients with primary hypothyroidism whereas no consistent elevation was found in patients with Graves' disease.

In 1956, Adams and Purves [6] first described the presence of a thyroid stimulator in the serum from patients with hyperthyroidism which differed from thyrotropin (TSH). This agent is now called long-acting thyroid stimulator, or LATS. Initially this substance was distinguished from pituitary thyrotropin by its longer duration of action, but with further investigation other differences between these two substances have been demonstrated.

METHOD OF ASSAY

The bioassay of TSH and LATS were performed according to the method of McKenzie [7] with minor modifications. Female albino Swiss-Webster mice weighing 15 to 20 grams were used. The mice were placed on the Remington low iodine diet [8] for at least two weeks (Table 8–III). At the end of this period of time, each mouse is injected, intraperitoneally, with 8 μCi of iodide ^{131}I. Four hours later, 15 μg of L-thyroxine, dissolved in a minimal amount of 0.1N sodium hydroxide, is injected intraperitoneally. Both substances are injected in a volume of 0.5 ml saline. Following two days of equilibration, the assay is performed on the morning of the fourth day.

When the preparation is complete either TSH standard, 0.9% saline solution or serum from patients is injected intravenously into the tail vein in a volume of 0.5 ml. Six mice are routinely used for each substance or dose tested and the effect is assessed by the measurement of the release of the prelabeled thyroid hormone into the circulation. Blood for determination of radioactivity is obtained by puncture of the retro-orbital area and collected in a heparinized capillary pipette at zero, two and ten to twelve hours following injection of the test material. One hundred microliters of blood is diluted with 2 ml of physiologic saline solution and the radioactivity determined in an automatic, well-type scintillation detector calibrated to count ^{131}I at peak efficiency. A maximal assay response is observed two hours after

TABLE 8–III
PROCEDURE FOR PREPARATION OF MICE FOR BIOASSAY OF TSH AND LATS

Time	*Procedure*
Day 1–14	Remington low iodine diet
Day 15	Intraperitoneal injection of 8 μCi I^{131} followed in 4 hours with the intraperitoneal injection of 15 μg L-thyroxine
Day 16–17	Time of equilibration
Day 18	Day of assay

the intravenous injection of TSH, whereas the maximal response with serum containing LATS is observed in the sample obtained twelve hours following the injection of the test substance. TSH and LATS values are expressed in terms of the radioiodine content of the blood at two hours and at twelve hours, respectively, as a percentage of the zero hour count which is assigned a value of 100 percent. Since LATS has a longer duration of action than TSH, the content of radiiodine at twelve hours greater than that at two is indicative of LATS activity, whereas, a value greater at two hours than at twelve hours is indicative of TSH activity. The test serum is considered LATS-positive when the twelve-hour blood value is at least 100 percent greater than the two-hour sample. Other methods used to express the results include giving the absolute response obtained in the assay[9] and deducting the mean effect of a control injection;[10] various statistical methods are logarithmic transformations of the raw data with subsequent covariance analysis for the independent variability of the zero hour count;[10] and Student's test comparing difference between the effects of test serum and the chosen control solution have been used to assess the LATS bioassay data. Figure 8–2 illustrates the results obtained from the assay of normal saline, 0.15 mμ of USP Thyrotropin Reference Standard dissolved in 0.9% saline solution and serum from a patient with Graves' disease. The in-

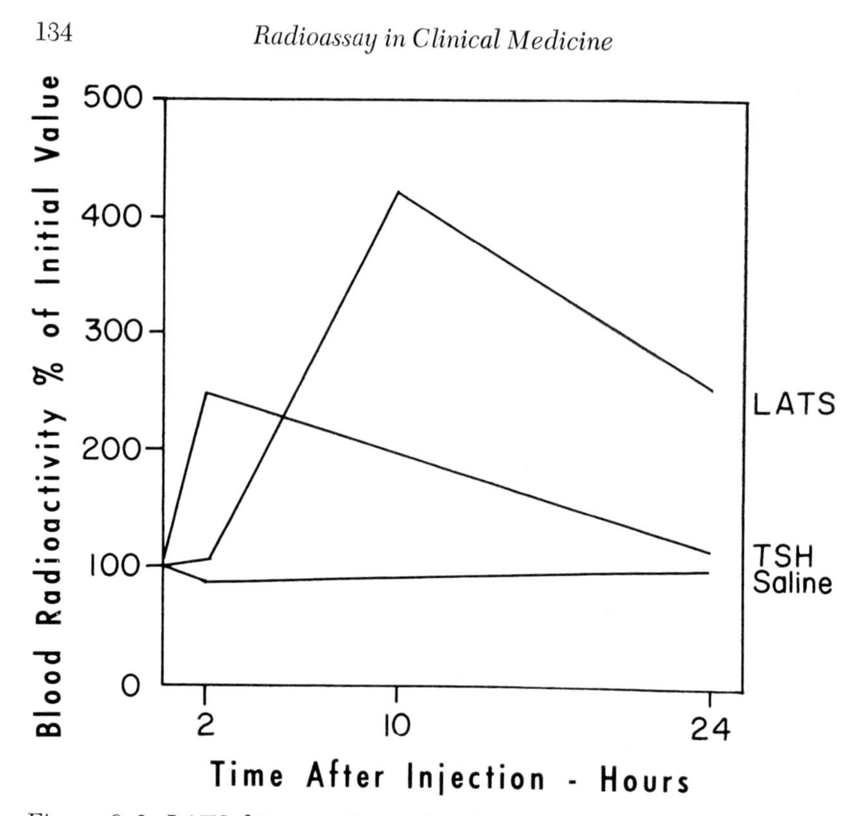

Figure 8–2. LATS bioassay. Note the differences in time of maximum radioactivity between LATS and TSH.

jection of saline did not after the level of circulating radioactivity. TSH reference standard elicited a maximum increase in radioactivity two hours after injection followed by a progressive fall-off thereafter. Similar results are obtained with the assay of serum from a patient with primary hypothyroidism. The serum from the patient with Graves' disease stimulates maximal release of preformed radioactive hormone twelve hours following the injection and is quite distinct from that obtained with TSH. It is this delayed effect that distinguished TSH from LATS.

DIFFERENCES BETWEEN TSH AND LATS

LATS was initially discovered because of the difference in the duration of action between TSH and LATS in the bioassay.

However, subsequently numerous additional differences have been described and reported. These differences can best be discussed under four major categories: biologic, antigenic, biochemical and site of origin.

BIOLOGIC DIFFERENCES

Following the early demonstration that LATS has a longer duration of action, studies were undertaken to determine the biologic half-lives of these two stimulators. In studies [11] performed on the rat, TSH has a half-life of ten to twenty minutes while LATS has a half-life of 7½ hours after injection. When TSH is detected in the serum of normal patients, the administration of exogenous thyroid hormone will suppress the circulating level of TSH, as would be expected from the negative feedback mechanism.[12] No such suppression of LATS is observed in patients with Graves' disease, indicating that LATS operates independent of the normal feedback mechanism. Clinically, the thyroid in patients with Graves' disease can neither be stimulated with TSH administration nor suppressed with exogenous thyroid hormone.

ANTIGENIC DIFFERENCES

In an early attempt to identify LATS, antiserum of bovine [13,14] or human TSH [15] failed to neutralize the hormonal activity of LATS, whereas the same antiserum completely inactivated TSH. Subsequently, Dorrington and Munro [16] demonstrated that LATS-IgG antiserum inactivated LATS but was ineffective against TSH.

BIOCHEMICAL DIFFERENCES

TSH can be separated from serum IgG (gammaglobulins) by procedures such as Adams' and Kennedy's fractional precipitation with acetone [17] or by the ethanol-salt percolation of Bates et al.[18] No TSH activity is found in serum containing only LATS. Fractionation of plasma containing LATS or TSH by the method of gel filtration showed that LATS was recovered in the 7 S fraction while TSH was recovered in the 4 S fraction.[19,20] Evidence indicates differing distribution coefficients in the gel. Digestive and proteolytic enzymes such as papain and subtilopeptidase destroyed thyrotropin. No such adverse effect was observed by incubating these enzymes with LATS.[20] In fact, Meek et al.[21] demonstrated that fragments obtained following papain digestion of LATS retained thyroid-stimulating properties that were TSH-like in nature and this activity could then be abolished by anti-human IgG antibody.

DIFFERENCES IN SITE OF ORIGIN

Thyrotropin has repeatedly been detected in pituitary glands with bioassay and immunoassay. LATS has been repeatedly detected in the blood of patients with Graves' disease but never in pituitary glands.[23] In addition, LATS has been demonstrated in patients developing hyperthyroidism following hypophysectomy.[25] This data would indicate that the anterior pituitary is the unlikely site of origin for LATS which therefore must arise from a different site. The knowledge that LATS is an immune globulin led investigators to study the lymphocyte as the possible source of this thyroid-stimulating globulin. McKenzie[26] and Miyai[27] have reported on the recovery of LATS from the culture medium of phytohemagglutinin-stimulated lymphocytes obtained from patients with LATS positive serum, thus suggesting that the lymphocyte may serve as the origin of this immunoglobulin.

CLINICAL SIGNIFICANCE OF LATS

LATS is not identified in the serum of normal patients. It is found specifically in patients with Graves' disease. Classic Graves' disease consists of goiter, thyrotoxicosis and exophthalmos. The greatest titers of LATS is found in patients with Graves' disease with accompanying pretibial myxedema. Euthyroid patients with or without goiter but with infiltrative ophthalmopathy also have LATS circulating in their blood.[28] The explanation for the absence of hyperthyroidism in these patients is that the thyroid gland is incapable of being stimulated. Therefore, in addition to the presence of LATS, a responsive thyroid gland must be present. LATS, however, is only found in unmodified whole serum in 60 percent of the patients with Graves' disease. Concentration of inactive serum from patients with Graves' disease will increase the positive yield by another 20 percent. The inability to detect LATS in all patients with Graves' disease may be due to inadequate sensitivity of the bioassay. The long-acting thyroid stimulator has not been isolated from patients with hyperfunctioning thyroid adenomas.[29]

A question may now be asked, is LATS the cause of the thyrotoxicosis? LATS is definitely capable of producing the thyrotoxic state if along with its presence there is a responsive gland. This is verified by the development of neonatal thyrotoxicosis.[36,37] This is a rare condition wherein a mother gives birth to a thyrotoxic infant. The mother either had a past history of Graves' disease or is thyrotoxic during the pregnancy. LATS is identified in the serum of both mother and infant. The clinical condition is the infant is usually self-limited followed by permanent remission and associated with a rapid decline of the LATS level in the child. Presumably, the condition is due to the transplacental passage of LATS from the mother to the fetus. The knowledge that LATS persists in the serum following the treatment[29,30] of the thyrotoxic state does not detract from the possibility that LATS is the cause of the thyrotoxicosis. Present-day treatment consists primarily in the management of the symptoms by direct attack on the thyroid with surgery, radioiodine or goitrogenic agents. None of these methods attempt to remove the specific cause of the condition but rather to control the synthesis and release of thyroid hormone.

Since LATS has also been identified in the serum of patients who are euthyroid but have infiltrative ophthalmopathy, the question of LATS as the cause of the ophthalmopathy arises. Albert[31] reported exophthalmos developed in the Atlantic minnow after injection of anterior pituitary extracts. Subsequently, Dobyns separated from crude anterior pituitary extracts two fractions, exophthalmos-producing substance (EPS) and purified thyrotropin.[32] In addition, Dobyns[33] found that injecting the serum of patients with infiltrative ophthalmopathy into the Atlantic minnow also produced striking exophthalmos. Initially, it was suggested that EPS and LATS were identical. Since then it has been shown that EPS was extractable from the anterior pituitary while LATS was not, thereby providing initial data contradicting the early hypothesis of similarity of these two substances. Pimstone et al.[34] performed parallel assays of thyrotropin (TSH), long-acting thyroid stimulator (LATS) and exophthalmos-producing substance (EPS) in serum of patients in various stages of treated and untreated Graves' disease with and without exophthalmos and failed to identify any similarity of these substances. The highest levels of LATS are found in patients with severe ophthalmopathy and pretibial myxedema. LATS, also, persists in the serum following successful management of the ophthalmopathy or the pretibial myxedema.

In summary, LATS is known to be found only in patients with Graves' disease and is an immunoglobulin IgG. It acts directly on the thyroid gland and not through the stimulation of any pituitary factor. It directly stimulates iodine incorporation, hormonal synthesis and release. In addition, LATS produces similar responses as TSH when its effect is studied on the intermediary metabolism of thyroid tissue *in vitro*.[35] Considerable evidence has been accumulated in identifying the activity of LATS as a property of specific IgG molecules. Unequivocal proof that LATS is an antibody will depend on the demonstration of an immunologic reaction between a specifically identified antigen and LATS.

The thyroid gland contains both soluble and particulate subcellular fractions capable of neutralizing LATS.[38,39] However, immunization of experimental animals with thyroid extracts has failed to reproduce the disease. One approach is to consider Graves' disease an autoimmune disease along with Hashimoto's thyroiditis and primary thyroid failure. Clinical observations suggest that these three thyroidal diseases may be interrelated; Graves' disease and Hashimoto's thyroiditis have been known to coexist and untreated thyrotoxicosis can develop into hypothyroidism. Therefore, the possibility exists that they may be different stages of the same disease process. The histologic picture in thyroiditis and Graves' disease is typical of the delayed hypersensitivity-type reaction. In addition, antibodies to thyroglobulin and thyroid cytoplasmic antibody are common to all these conditions. LATS, however, is only found in Graves' disease and perhaps can be viewed as a marker of active disease. These same disorders of the thyroid also demonstrate gastric microsomal antibodies. An interesting hypothesis is that these entities represent a genetically determined defect in immunological tolerance.[38] This may be supported by the observation that the extended administration of triiodothyronine [39] to patients with active Graves' disease is associated with the hypertrophy of the cervical lymphoid structure and formation of an excess of immunoglobulins IgG, LATS and IgM. While the puzzle remains unsolved at this time, certain facts indicate an autoimmune disease of the thyroid wherein LATS is a marker of Graves' disease.

REFERENCES

1. Salter, W.T.: *The Endocrine Function of Iodine.* Cambridge, Mass., Harvard University Press, 1940.
2. Schreiber, V., Rybak, M., Eckertova, A., Koci, J., Jirge, V., Franc, A., and Kmentova, V.: Isolation of a hypothalamic peptide with TRF (thyrotropin releasing factor) activity *in vitro. Experientia, 18:*338, 1962.
3. McKenzie, J.M.: Bio-assay of thyrotropin in man. *Physiol Rev, 40:* 398, 1960.
4. Utiger, R.D.: Radioimmunoassay of human plasma thyrotropin. *J Clin Invest, 44:*1277, 1965.
5. Lemarchard-Beraud, T., and Vannotti, A.: A radioimmunoassay for the determination of thyroid stimulating hormone. *Experientia, 21:*353, 1965.
6. Adams, D.D., and Purves, H.D.: Abnormal responses in the assay of thyrotropin. *Proc Univ Otago Med School, 34:*11, 1956.
7. McKenzie, J.M.: The bioassay of thyrotropin in serum. *Endocrinology, 63:*372, 1958.
8. Ungar, F., and Halberg, F.: *In vitro* explanation of a circadian rhythm in adrenocorticotropic activity of C mouse hypophysis. *Experientia, 19:*158, 1963.
9. Burke, G.: Failure of immunologic reaction of long-acting thyroid

stimulator (LATS) to thyroid components and demonstration of a plasma inhibitor of LATS. *J Lab Clin Med, 69:*713, 1967.

10. McKenzie, J.M., and Williamson, A.: Experience with the bioassay of the long acting thyroid stimulator. *J Clin Endocrinol, 26:*518, 1966.
11. Adams, D.D.: A comparison of the rates at which thyrotropin and the human abnormal thyroid stimulator disappear from the circulating blood of the rat. *Endocrinology, 66:*658, 1960.
12. Adams, D.D.: The presence of an abnormal thyroid stimulating hormone in the serum of some thyrotoxic patients. *J Clin Endocrinol, 18:*699, 1958.
13. Cline, M.J., Selenkow, H.A., and Brood, M.S.: Inhibition of thyrotropic activity by immunologically specific antiserum. *Endocrinology, 67:*273, 1960.
14. McKenzie, J.M., and Fishman, J.: Effects of antiserum in bioassay of thyrotropin and thyroid activator of hyperthyroidism. *Proc Soc Exp Biol Med, 105:*126, 1960.
15. Adams, D.D., Kennedy, T.H., Purves, H.D., and Sirett, N.E.: Failure of TSH antisera to neutralize long-acting thyroid stimulator. *Endocrinology, 70:*801, 1962.
16. Dorrington, K.J., and Munro, D.S.: Immunological studies on the long-acting thyroid stimulator. *Clin Sci, 28:*165, 1965.
17. Adams, D.D., and Kennedy, T.H.: Evidence of normally functioning pituitary TSH secretion mechanism in a patient with a high blood level of long-acting thyroid stimulator. *J Clin Endocrinol, 25:*571, 1965.
18. Bates, R.W., Garrison, M.M., and Howard, T.B.: *Extraction of* thyrotropin from pituitary glands, mouse pituitary tumors and blood plasma by percolation. *Endocrinology, 65:*7, 1959.
19. McKenzie, J.M.: Fractionation of plasma containing the long-acting thyroid stimulator. *J Biol Chem, 237:*3571, 1962.
20. McKenzie, J.M.: Enzymic hydrolysis of thyrotropin and the long-acting thyroid stimulator. *J Clin Invest, 42:*955, 1963.
21. Meek, J.C., Jones, A.E., Lewis, V.J., and Vanderlaan, W.P.: Characterization of the long-acting thyroid stimulator of Graves' disease. *Proc Natl Acad Sci, 52:*342, 1964.
22. Kriss, J.P., Pleshakov, V., Rosenblum, A., and Chien, J.R.: Studies on the formation of long-acting thyroid stimulator globulin (LATS) and the alteration of its biologic activity by enzymatic digestion and partial chemical degradation. *Current Topics in Thyroid Research.* New York, Academic Press, Inc., 1965.
23. Miyai, P.W., and Munro, D.S.: Observations on the stimulation of thyroid function in mice by the injection of serum from normal subjects and from patients with thyroid disorders. *Clin Sci, 23:*463, 1962.

24. Furth, E.D., Becker, D.V., Ray, B.S., and Kane, J.W.: Appearance of unilateral infiltrative exophthalmos of Graves' disease after the successful treatment of the same process in the contralateral eye by apparently total surgical hypophysectomy. *J Clin Endocrinol, 22:*518, 1962.
25. Christensen, L.K., and Binder, V.A.: A case of hyperthyroidism developed in spite of previous hypophysectomy. *Acta Med Scand, 172:*285, 1962.
26. McKenzie, J.M., and Gordon, J.: The origin of the long-acting thyroid stimulator. In Cassano, C., and Andreoil, M. (Eds): *Current Topics in Thyroid Research,* Proceedings of the 5th International Thyroid Conference, Rome. New York, Academic Press, 1965.
27. Miyai, K., Fukuchi, M., and Kumahara, Y.: LATS production by lymphocyte culture in patients with Graves' disease. *J Clin Endocrinol, 27:*855, 1967.
28. Liddle, G.W., Heyssel, R.M., and McKenzie, J.M.: Graves' disease without hyperthyroidism. *Am J Med, 39:*845, 1965.
29. McKenzie, J.M.: Review. Pathogenesis of Graves' disease: Role of the long-acting thyroid stimulator. *J Clin Endocrinol, 25:*424, 1965.
30. Volpe, R., Desbrats-Schonbaum, M.L., Schonbarum, E., Row, V.V., and Ezrin, C.: The effect of radioablation of the thyroid gland in Graves' disease with high levels of long-acting thyroid stimulator (LATS). *Am J Med, 46:*217, 1969.
31. Albert, A.: The experimental production of exophthalmos in fundulus by means of anterior pituitary extracts. *Endocrinology, 37:*389, 1945.
32. Dobyns, B.M., and Steelman, S.L.: The thyroid stimulating hormone of the anterior pituitary as distinct from the exophthalmos producing substance. *Endocrinology, 52:*705, 1953.
33. Dobyns, B.M., and Wilson, L.A.: An exophthalmos producing substance in serum of patients suffering from exophthalmos. *J Clin Endocrinol, 14:*1393, 1956.
34. Pimstone, B.L., Hoffenberg, R., and Black, E.: Parallel assays of thyrotropin, long-acting thyroid stimulator, and exophthalmos-producing substance in endocrine exophthalmos and pretibial myxedema. *J Clin Endocrinol, 24:*976, 1965.
35. Scott, T.W., Good, B.F., and Ferguson, K.A.: Comparative effects of long-acting thyroid stimulator and pituitary thyrotropin on the intermediary metabolism of thyroid tissue *in vitro. Endocrinology, 79:*949, 1966.
36. Sunshine, P., Kusumoto, H., and Kriss, J.P.: Survival time of circulating long-acting thyroid stimulator in neonatal thyrotoxicosis: Implication for diagnosis and therapy of the disorder. *Pediatrics, 36:*869, 1965.
37. Hoffman, M.J., Hetzel, B.S., and Manson, J.: Neonatal thyrotoxi-

cosis: Report of three cases involving four infants. *Aust Ann Med, 15:*262, 1966.

38. Irvine, W.J.: Thyroid auto-immunity as a disorder of immunological tolerance. *QJ Exp Physiol, 49:*324, 1964.
39. Mahaux, J.E., Chamla-Soumenkoff, J., Delcourt, R., Nagel, N., and Levin, S.: The effect of triiodothyronine on cervical lymphoid structures, thyroid activity, IgG and IgM immunoglobulin level and exophthalmos in Graves' disease. *Acta Endocrinol, 61:*400, 1969.

CHAPTER 9

ERYTHROPOIETIN BIOASSAY

ROBERT M. DONATI AND NEIL I. GALLAGHER

INTRODUCTION

IN THE LAST TWO DECADES, we have witnessed a rapid increase in the fundamental knowledge concerning erythropoiesis. For many years the constancy of the circulating red blood cell mass has been noted, but the factors involved in maintaining a stable red blood cell concentration by regulating the rate of erythropoiesis were unknown. Experiments that subjected animals or man to atmospheric hypoxia demonstrated increase in erythropoiesis, while hyperoxia due to plethora, produced by the transfusion of red blood cells, was shown to decrease erythropoiesis. This led to the concept that the rate of erythropoiesis was controlled by the marrow oxygen tension. Experiments by Grant and Root,[1] however, measuring marrow oxygen concentration, could not substantiate the hypothesis that rate of erthyropoiesis was regulated by marrow oxygen tension.

In the 1950s', the original hypothesis put forward by Carnot and Deflandre,[2] that erythropoiesis was controlled by a humoral factor, elaborated outside the marrow and secreted into the blood in response to hypoxia, was reevaluated. Reissman[3] provided major impetus to the study of the humoral control of erythropoiesis by the demonstration that subjecting one parabiotic rat to hypoxia, which produced a blood oxygen saturation of 63 percent, increased erythropoiesis in the partner rat breathing room air and having a blood oxygen saturation of 97 percent. The only explanation for this observation was transmission to the rat with normal oxygen saturation of a humoral factor that increased erythropoiesis. Stohlman and his coworkers[4] subsequently had an opportunity to study a patient with a patent ductus arteriosis and reversal of flow. As a result of this cardiac abnormality, the patient exhibited a normal oxygen saturation above the diaphragm and a decreased oxygen saturation below the diaphragm. Erythroid hyperplasia demonstrated in the sternal marrow, which was normally oxygenated, provided further evidence that erythropoiesis was not under the control of local marrow oxygen tension but rather controlled by a humoral factor produced below the diaphragm. Subsequently, within a short time period, Borsook and his associates,[5] Erslev,[6] Gordon and his associates [7] and Jacobson's group [8] were all able to confirm Carnot and Deflandre's hypothesis by injecting plasma from anemic animals into normal animals and observing an increase in blood reticulocytes or the incorporation of radioiron into red blood cells. The plasma factor that increased erythropoiesis had been termed "hemopoietine" by Carnot and Deflandre; however, as work proceeded, it appeared to be involved exclusively with red blood cell production, and the term erythropoietin was adopted.

ORIGIN AND ACTION OF ERYTHROPOIETIN

The result of numerous investigations over the last fifteen years has been the partial characterization of the site of production and mechanism of action of erythropoietin. On the basis of early extirpation experiments, Jacobson and associates [9] concluded that erythropoietin was either produced or activated in the kidney. Many investigators have confirmed the importance of the kidney and have demonstrated that the kidney is the primary source of erythropoietin production. However, other investigations [10] have suggested that the kidney may not be the sole source of erythropoietin in all species. Gordon's group [11] have suggested an erythropoietin precursor generating system which is analogous to the renin angiotensin generating system. In this system, an erythropoietin precursor is produced in the liver, secreted into the plasma and is activated by a renal activating substance.

Although the consensus is that the kidney is the primary source of erythropoietin, the specific cellular site of production is not known. The renal cortex, the renal medulla, both the cortex and the medulla, the kidney tubules or renal vasculature have all been suggested as the site of erythropoietin production. Alternatively, these sites may respond to a messenger substance from another site such as the juxtaglomerular cells. The juxtaglomerular cells were first suggested as the site of erythropoietin production by Osnes in 1954.[12] This view has been held by other investigators.[13,14] The parallel association of erythropoietin production and juxtaglomerular cell granularity led these investigators to suggest that the juxtaglomerular cells may secrete erythropoietin. However, the evidence for the secretion of erythropoietin by the juxtaglomerular cells is indirect. Studies in this and other laboratories have been unable to document the production of erythropoietin by the juxtaglomerular cell. However, stimuli for erythropoiesis have been dissociated from the stimuli for renin secretion.[15,16]

While the cellular site of origin is in doubt, the overwhelming evidence supports the concept that erythropoietin is produced by the kidney in response to the level of oxygen saturation in the blood perfusing the kidney. Thus, with diminished renal oxygenation, there is augmentation of the production and release of erythropoietin and a subsequent increase in red blood cell production.

Evidence from a number of investigators supports the hypothesis that the erythropoietin produced in the kidney exerts its influence at the stem cell level of the marrow, causing the differentiation of these cells into erythrocytic elements.[17-22] Studies utilizing the injection of antisera to erythropoietin into polycythemic mice pre-treated with erythropoietin support this view. Although erythropoietin exerts an effect on the stem cells, this is apparently not its only action. Erythropoietin also affects erythroid proliferation, maturation and release of erythrocytes from the marrow. This has led Linman[23] to postulate two erythropoietins; one affecting the rate of erythrocyte differentiation and the other determining the number of mitoses which the erythrocyte precursors undergo during maturation.

The present concept is that erythropoiesis is regulated in the following manner. Erythropoietin production occurs mainly in the kidney with approximately 10 percent occurring from other sources. Production is thought to be regulated primarily by the balance of oxygen supply and demand with the endocrine glands exerting their action as a secondary influence on metabolic reactions. Thus, the augmented erythropoiesis which occurs secondary to thyroid hyperactivity would be the result of an imbalance in tissue oxygen supply and tissue oxygen demand producing relative hypoxia in the kidney with the production and secretion of erythropoietin. A small amount of the erythropoietin is excreted in the urine, some of it being inactivated and inhibited by urinary proteins. The possibility that there is another system in addition to the erythropoietic stimulatory system which acts to modulate broad swings in red cell production by means of inhibiting erythropoiesis has been the subject of considerable investigation,[24,25] but a definitive demonstration of such a system has not as yet been forthcoming. While erythropoietin exerts its main action on the stem cells of the bone marrow, more peripheral actions also occur, including further effects on red blood cell proliferation and maturation and the release of reticulocytes from the bone marrow. The developed red blood cells contribute to a feedback through their influence on oxygen supply, perhaps the destruction of the red blood cells and the release of hemoglobin from these cells represents another controlling feedback mechanism. Thus, we have a system modulating red blood cell production in which the major control is oxygenation and oxygen balance, affecting the kidney so that diminished oxygen supply stimulates the production of erythropoietin. Erythropoietin acts on both the stem cell to produce increased red blood cell precursors and on the erythroid marrow to produce erythroid maturation and reticulocyte release. The subsequent increase in red cell mass leads to an increase in oxygen supply and the system is modulated by this feedback. However, the existence of specific erythropoietic inhibitors cannot be excluded from such a system at this time.

BIOASSAYS

As in all hormonal systems, the *sine qua non* for effective determination of the activity of the hormone and establishment of some relationship to clinical systems depends on the assay method available for determination of hormonal activity. Since erythropoietin has only recently been purified in miniscule amounts, an immunoassay system which is reproducible has yet to be developed. Thus, radiobioassays remain the mainstay for indirectly detecting erythropoietin in body fluids and the tissue extracts and for studying the rate of production and utilization of the hormone as well. The early systems used to bioassay erythropoietin utilized primarily the determination of the reticulocyte response following the administration of a test dose of plasma or test urine. The assay of erythropoietin became easier and more accurate with the measurement of the utilization of an injected dose of radioiron by the red blood cell as the index of erythropoiesis.

Basic to all bioassays which have been developed for the assessment of erythropoietin is the production of a low baseline level of erythropoiesis in the assay animal, the administration of the test substance and the subsequent assessment of the erythropoietic response of the assay animal. Diminished erythropoiesis has been produced in the test animals either by starvation,[26] hypophysectomy,[27] or by increasing the red cell mass either by transfusion-induced plethora [28] or exposure to a hypoxic stimulus.[29,30] All of these methods reduce erythropoiesis by suppression of endogenous erythropoietin and increase the sensitivity of the test animals to the administration of test samples containing erythropoietin. Following the suppression of endogenous erythropoietin production by one of the means detailed above, the assay animal is injected with test sera or urine and the resultant change in erythropoiesis determined.

The method of measuring the alteration in the rate of erythropoiesis generally has been the administration of a tracer dose of ^{59}Fe and the subsequent determination of its incorporation after a standard time, into the circulating red blood cell mass. Thus, one can determine whether or not the test substance is erythropoietically active or erythropoietically inactive.

Our laboratory uses discontinuous hypobaric hypoxia or hypoxic hypoxia to produce plethora in the assay animal which is a CF #1 mouse. Hypobaric hypoxia is produced in a chamber maintained by means of a vacuum at 0.5 atmosphere pressure. Hypoxic hypoxia is produced by means of a mouse cage covered with a silicone rubber membrane. Because of the difference in permeability, the partial pressure difference needed to drive carbon dioxide out of the cage is less than one-fifth that necessary to supply oxygen to the animals. As a consequence, carbon dioxide levels are maintained within a tolerable range. The level of hypoxia is proportional to the number of animals in the cage.[30] In the bioassay of erythropoietin utilizing hypobaric hypoxia the mice are subjected to 0.5 atmosphere for two weeks. They then are moved from the hypobaric chamber and returned to normal room air. The second day following the removal from the chamber, the animal is given an injection of test material. On the fourth day following removal from the chamber, the animal is given an intravenous injection of a tracer dose of 1 μCi of ^{59}Fe/1 μg ^{56}Fe citrate. Eighteen or forty-eight hours following the injection, the animal is killed and the incorporation of radioiron into the red blood cells is determined. An example of the results of this sort of manipulation is shown in Table 9–I which compares blood volumes, hematocrits and red blood cell volume and RBC ^{59}Fe incorporation of normal mice and mice maintained in hypobaric hypoxia. The RBC radioiron incorporation of normal mice was approximately 24 percent. Following fourteen days in a hypoxic environment, their packed cell volume had increased to 62 vol percent, whereas, the radioiron incorporation remained at approximately 23 percent. Ninety-six hours following removal from the hypobaric hyperoxia, the packed cell volume remained increased, however, the percent of the injected dose of radioiron incorporated into the erythrocytes had diminished to 3.3 percent, thus providing a sensitive animal for the assay of erythropoietin.

The erythropoietic response to erythropoietin in both normal and hypoxic mice handled in this manner is presented in Table 9–III. This data demonstrates the increased sensitivity of the mouse which is prepared with two-weeks hypobaric hypoxia and subsequently returned to normal atmospheric conditions to the same dose of erythropoietically active material.

In the assay of urinary erythropoietin, we have developed a somewhat different procedure which is a modification of that of

TABLE 9–I

BODY WEIGHT, RED BLOOD CELL MASS, PACKED CELL VOLUME AND BLOOD VOLUME OF NORMAL AND HYPOXIC MICE

	Number of mice	*Body Weights gm ± SDM*	*Packed Cell Volume % ± SDM*	*Red Cell Mass ml /100gm ± SDM*	*Blood Vol/ 100 gm ± SDM*
Normal	13	24.1 ± 3 *	46 ± 4.3	0.7 ± 0.1	6.8 ± 0.6
0.5 atm 14 DAYS	12	22.8 ± 2 * 22.9 ± 2.2 †	59 ± 3.2	1.2 ± 0.4	8.7 ± 1.1
0.5 atm 14 DAYS 1.0 atm 96 HOURS	9	19.9 ± 1.1 * 18.7 ± 1.9 † 22.4 ± 1.1 ‡	62 ± 3.7	1.10 ± 0.3	7.9 ± 0.5

* Initial.
† 14 Days at 0.5 atm
‡ 14 Days at 0.5 atm and 96 hours at 1.0 atm.

TABLE 9-II
EIGHTEEN-HOUR ^{59}Fe RED BLOOD CELL INCORPORATION IN NORMAL AND HYPOXIC MICE

MICE	*NUMBER*	*PACKED CELL VOL % ± SDM*	*^{59}Fe INCORPORATION % ± SDM*
Normal	15	44 ± 1.9	24.1 ± 7.4
0.5 atm 14 days	19	62 ± 3.1	23.9 ± 5.5
0.5 atm 14 days 1.0 atm & 96 hours	16	63 ± 4.2	3.3 ± 1.2

Adamson.[39] A twenty-four-hour urinary sample for erythropoietin is collected by means of a simply contrived apparatus consisting of a picnic cooler, dry ice, a polyethylene bottle and a funnel, which maintains the urine at dry ice temperature during the period of collection. Collection of urine in this manner prevents the loss of erythropoietic activity which characteristically occurs at higher temperatures.

Following the collection of the twenty-four-hour samples in dry ice, the sample is dialyzed against tap water and then against distilled water. The sample is then lyophilized and reconstituted in 100 ml of distilled water so that 1 ml of reconstituted concentrate equals 1 percent of the daily urine output. Following preparation, the urine is assayed in the same sort of exhypoxic plethoric mouse bioassay detailed above. The only difference is that on days three and four, the bioassay animal receives a total of four injections of 1 ml of urinary concentrate so that 4 percent of the total daily urinary output of erythropoietin is assayed. This sort of assay provides for the measurement of levels of erythropoietin in the urine which are within the normal range and those which are depressed below normal, and provides greater sensitivity and accuracy than the plasma assay.

A possible alternative to the urine assay which may be as sensitive has been recently investigated in our laboratory. Plasma from normal male donors was bioassayed in exhypoxic mice. The mice were rendered plethoric by discontinuous exposure to 0.5 atm for two weeks. On the second and third day following removal from the hypobaric chamber, each mouse was injected with one half of the amount of plasma recorded on horizontal axis. Iron 59 was injected on the fifth posthypoxic day and a

TABLE 9–III

ERYTHROPOIETIC RESPONSE TO ERYTHROPOIETIN IN NORMAL AND HYPOXIC MICE

MICE	*NUMBER*	*MATERIAL*	*PACKED CELL VOLUME* *% ± SDM*	*% 18 HOURS ^{59}Fe* *INC. ± SDM*
Normal	15	Not injected	44.0 ± 1.9	24.1 ± 7.4
	8	Sheep erythropoietin 5 cobalt units	46.0 ± 2.9	40.4 ± 5.7
	8	Aplastic anemia plasma 0.4 ml	47.0 ± 1.7	46.3 ± 5.0
0.5 atm 14 Days and 1.0 atm 96 Hours	16	0.2 ml saline	63.0 ± 4.2	3.3 ± 1.2
	3	Sheep erythropoietin 5 cobalt units	57.0 ± 2.0	23.7 ± 9.4
	16	Aplastic anemia plasma 0.4 ml	64.6 ± 3.6	37.4 ± 5.3

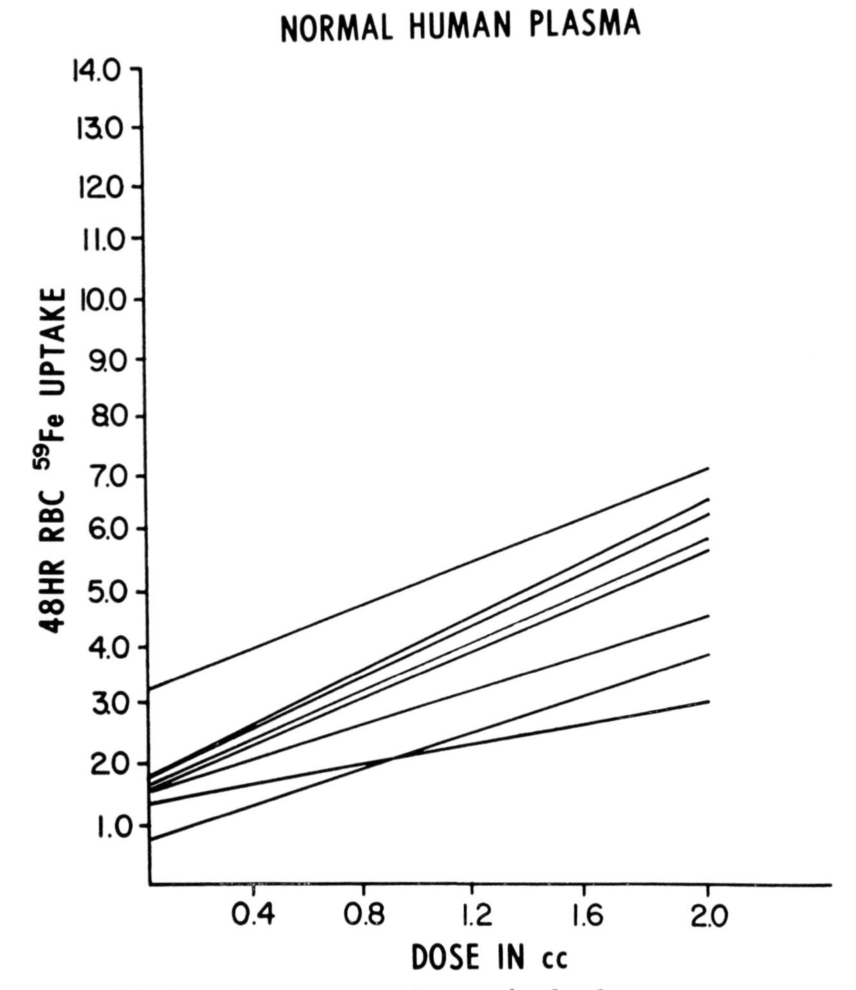

Figure 9–1. Dose Response curve. See text for details.

forty-eight-hour red blood cell ^{59}Fe uptake used as the measure for comparison of the effects of the plasma injections on cell production. These dose response curves represent calculated best fit lines. Figure 9–1 reveals a summation of this data. Each line represents the calculated best fit line derived from the mean of five assays at each level of plasma for an individual patient. It is our belief that the need for a sensitive and reproducible bioassay continues, and that standardization of this technique may provide a method for the assay of "normal levels" of erythropoietin.

That the slopes actually represent erythropoietin in the normal plasma remains open to question, but in three additional experiments, we have partially "inactivated" normal plasma by storage for three to seven days at 20° C or room temperature. The slope of the curve is less, but in no instance has the plasma lost all activity. In addition, further proof that the activity measured is truly erythropoietin may be demonstrated by inactivation with an antierythropoietin.

In several instances, this plasma assay method has been helpful in demonstrating that erythrocytosis was secondary to elevated erythropoietin; however, it has not helped us in establishing the diagnosis of polycythemia vera. Finally, an unexpected reticulocytopenia in a patient with hemochromatosis treated by serial phlebotomy was found to be due to failure to produce erythropoietin rather than marrow insensitivity.

CLINICAL APPLICATION

The clinical utility of the erythropoietin bioassay has been a subject of considerable investigation. Plasma and urine erythropoietin titers from patients with diverse types of anemia have been measured. In general, elevated levels have been found in posthemorrhagic, vitamin deficiency, hemolytic and refractory anemias.[31] However, in the situation of chronic renal disease with anemia, erythropoietin levels are found to be only slightly increased in approximately 50 percent of the patients and normal or diminished in the remainder, thus suggesting that the inability to elaborate erythropoietin in response to an anemic stimulus plays a role in the pathogenesis of this anemia.

Perhaps the most important area in which erythropoietin levels can be utilized in clinical diagnosis is in the assessment of the etiology of the polycythemic syndromes; and, indeed, the erythropoietin system has been utilized as the basis for the classification of pathophysiologic states characterized by erythrocytosis. The remainder of the discussion will revolve about the role of erythropoietin and its measurement in the differentiation of the etiologies of the polycythemic states.

TABLE 9–IV

CLASSIFICATION OF POLYCYTHEMIAS

I. PRIMARY POLYCYTHEMIA (IDIOPATHIC) PROLIFERATION OF ABNORMAL STEM CELL WHICH DOES NOT REQUIRE ERYTHROPOIETIN FOR DIFFERENTIATION OR MATURATION. DIMINISHED EXCRETION OF ERYTHROPOIETIN.

II. SECONDARY ERYTHROCYTOSIS—DUE TO INCREASED PRODUCTION OF ERYTHROPOIETIN.

- A. Erythropoietin production in response to hypoxemia.
  - 1. Arterial oxygen desaturation due to living at high altitudes, diseases of ventilation or gaseous diffusion or from the presence of a right to left cardiac shunt.
  - 2. Altered hemoglobin oxygen affinity.
    - a. Genetic (hemoglobin Kempsey, Yakima, Ranier, Ypsi, and Chesapeake. Congenital methemoglobinemia).
    - b. Acquired (methemoglobinemia, sulf-hemoglobinemia, cobalt intoxication?)
- B. Erythropoietin production autonomous without a hypoxemic physiologic stimulus.
  - 1. Renal diseases (non-neplastic). See Table 9–V.
  - 2. Neoplasms. See Table 9–V.
- C. Erythropoietin production due to unknown mechanism of endocrine interaction.
  - 1. Cushing's Syndrome.
  - 2. Testosterone.
  - 3. Masculinizing ovarian tumors—luteomas.

Polycythemia is defined as a condition in which the absolute number of red blood cells in the blood is increased. This excludes the cases of stress polycythemia which represent only a decrease in the plasma volume and a consequent increase in the hematocrit with no increase in total number of red cells. These conditions would better be called stress hypovolemia.

Table 9–IV presents classification of polycythemias modified from the classification of Modan in 1965 [32] and Levin and Alperin in 1968.[33] This classification separates polycythemia vera and the secondary polycythemias. The latter are either anoxemic or hormonal dependent. Here we have divided polycythemias or erythrocytosis into those which are essential or idiopathic such as polycythemia vera, where there is an abnormal stem cell which is autonomous and not dependent upon erythropoietin for erythrocytogenesis. The other groups are those which are due to increased erythropoietin production and these are the erythrocytoses secondary to hypoxemia, hemoglobin abnormalities, diseases of the kidney, and neoplastic diseases, and have in common a diminished oxygen supply and increased erythropoietin production.

Polycythemias which result from augmented erythropoietin production may also be divided into those in which the increased erythropoietin levels are responsive to a physiological stimulus such as hypoxia, and those in which no physiological stimulus appears to be present for the increased erythropoietin levels. In the former case, normal stimuli increase erythropoietin production which increases erythropoiesis until the oxygen supply matches the demand and a new homeostasis is established with an increase in the red cell mass. In the latter, erythropoietin may be produced autonomously due to renal tumors or other neoplasms, or abnormalities. Polycythemia vera has been termed primary polycythemia since no cause for the increased erythropoiesis is apparent, while other polycythemias have been termed secondary polycythemias due to the presence of a recognized cause for the polycythemia and increased erythropoietin levels.

We might digress to discuss the erythrocytosis which is found in association with renal abnormalities or tumors. These very interesting occurrences have been described with a variety of tumors and renal abnormalities.[34,35] The renal abnormalities and tumors associated with erythrocytosis are indicated in the Table 9–V. These syndromes are characterized by the elaboration of erythropoietin presumably by the tumor itself or by the kidney in response to the renal abnormality.

Recently, Alexanian [36-38] and Adamson,[39,40] working independently, have utilized the urinary erythropoietin bioassay to characterize normal urinary erythropoietin excretion in normal man, patients with hypoxic erythrocytosis or erythrocytosis secondary to tumors or renal lesions.

TABLE 9–V
NEOPLASTIC CAUSES OF ERYTHROCYTOSIS

Renal Carcinoma	Wilms Tumor
Cerebellar Hemangioblastoma	Uterine Myomata
Hepatocellular Carcinoma	Adrenal Adenoma
Pheochromocytomas	Carcinoma of the Lung
Renal Adenoma	

Phlebotomy in normal male volunteers resulted in a progressive rise in urinary erythropoietin levels during the first week and maximum values were reached eight to twelve days after phlebotomy. Excretion remained elevated for one week and then declined as the red cell volume returned to normal range.[36-38] Similar data were derived from the studies of Adamson. A similarity between slopes of response over widely varying ranges in hematocrits was evident.

Under basal conditions, thirty normal and anemic males demonstrated an increase in the erythropoietin excretion exponentially with the degree of the anemia. In the thirteen patients who were bled, the erythropoietin excretion at the new red blood cell volume was similar to the excretion found in patients with comparable degrees of chronic anemia. In addition, in three patients with polycythemia vera, anemia was maintained for three to five months by repeated phlebotomy. These patients demonstrated a fall in erythropoietin excretion to within the range of that for patients with similar degrees of anemia from bone marrow failure. In normal man, a five-fold increase in urinary erythropoietin occurred after phlebotomy and was associated with a 2.5-fold increase in red cell production. In anemic patients, marked elevations in erythropoietin over a

100-fold range were usually associated with normal or low values for erythron iron turnover. In three of the five patients with polycythemia vera, following phlebotomy there was an elevation in erythron iron turnover of at least 20 percent. In the other two patients, there was no change in iron turnover.

With this baseline data on the normal urinary erythropoietic response and its relation to the red blood cell mass, patients with erythrocytosis were further evaluated. Values for urinary erythropoietin excretion in patients with polycythemia vera were lower than normal. However, there was some overlap with the normal range. In patients with hypoxic erythrocytosis or erythrocytosis secondary to tumors or renal abnormalities, the urinary erythropoietin levels were generally normal or elevated. The problem which now arises is how to differentiate the pathologic states which overlap the normal levels.

Studies which have been undertaken by Adamson and his associates [40] have clarified this problem to a degree. Phlebotomy of normal individuals raised the mean daily urinary erythropoietin excretion by several units. If the urinary erythropoietin levels were near normal in patients with hypoxia and erythrocytosis, or hypoxemia due to abnormal hemoglobins with an altered oxygen dissociation curve, phlebotomy resulted in an exaggerated increase in the daily urinary excretion of the hormone which distinguished these cases from normal individuals. The reason for this is unknown, but one may speculate that in these cases of erythrocytosis, the daily urinary excretion of hormone will vary depending on the environmental state and the daily level of oxygen supply and demand, so that at rest in a controlled environment, the hormone level may decline toward normal values. When the patients are phlebotomized, however, an increased ability to secrete erythropoietin, perhaps gained by previous prolonged periods of hypersecretion, may become apparent. A similar type of response has been demonstrated in patients with familial erythrocytosis, an inherited autosomal trait, as well as in patients with abnormal hemoglobins with an altered oxygen dissociation curve. This suggests that the erythrocytosis in these individuals is due to increased secretion which is not apparent.

In contrast to the situation with the hypoxic polycythemics, in patients with inappropriate erythrocytosis due to autonomous erythropoietin production, phlebotomy resulted in no further increase in urinary erythropoietin levels. This kind of polycythemia has been reported in patients with hypernephromas, Wilms tumor, hemangioblastomas, adenomas, fibromyxomas and adrenal adenomas. This type of response is unique and may reflect the fact that in many patients with malignancies, a hypoproliferative anemia is seen which is contributed to by depressed erythropoietin production. Although the response to phlebotomy was certainly depressed, it is likely that if the hematocrit had been lowered to more anemic levels, there would have been stimulation of erythropoietin production by the kidney. Nevertheless, the patterns of response in these patients support the concept of unregulated production of erythropoietin by neoplastic tissue and provide a means of differentiating polycythemic states.

Although it is infrequent that in cases of polycythemia vera one would need to make the differentiation since urinary erythropoietin levels in this instance are almost universally below normal, the response to phlebotomy was measured in these patients. Prior to the study, the excretion of erythropoietin was virtually immeasureable; however, with lowering of the hematocrits, six of the seven patients had an erythropoietin response. This response appeared similar in shape but of less magnitude than that seen in normal subjects. Following challenging by repeated phlebotomies, there was usually a step-wise increase in the excretion of erythropoietin. The physiologic situation in this disease resembles that of continuous autotransfusion where red cell production is independent of oxygen deprivation and erythropoietin stimulation. This interpretation is consistent with the inability of the oxygen-rich environment or hypertransfusion to suppress erythropoiesis in this disease. As has been suggested, red blood cell production is not under the regulation of erythropoietin and must be considered autonomous.

In summary, in the evaluation of patients who have erythrocytosis, the determination of the urinary levels of erythropoietin has provided a method of pathophysiologic classification and differentiation. In instances of polycythemia rubra vera, the twenty-four-hour urinary erythropoietin excretion is markedly depressed and virtually immeasurable. These patients may respond to phlebotomy by augmenting urinary production of erythropoietin; however, the response is less than normal. In contrast, the patient with hypoxic erythrocytosis may have an elevated urinary erythropoietin level, although it is frequently within the limits of normal. Following phlebotomy, the patient with hypoxic polycythemia demonstrates a grossly accelerated abnormal hyper-response to the phlebotomy with a dramatic increase in urinary erythropoietin levels. Furthermore, the patient who has tumorogenic erythrocytosis may have urinary erythropoietin levels which are initially elevated although sometimes within the limits of normal. However, following phlebotomy the patient's urinary erythropoietin excretion is minimally altered. Thus, the utilization of urinary erythropoietin bioassay enables the adequate evaluation and classification of the etiologies of the erythrocytosis in clinical situations.

REFERENCES

1. Grant, W.C.: The influence of anomia of lactating rats and mice on blood of their normal offspring. *Blood, 10*:334, 1955.
2. Carnot, P., and Deflandre, C.: Sur l'activite hemapoietique des differents organes au cours de la regeneration du sang. *Compt Rend Acad Sci, 143*:432, 1906.
3. Reissman, K.R.: Studies on the mechanism of erythropoietic stimulation in parabiotic rats during hypoxia. *Blood, 5*:372, 1950.
4. Stohlman, F., Jr., Rath, C.E., and Rose, J.C.: Evidence for a humoral regulation of erythropoiesis. Studies on a patient with polycythemia secondary to regional hypoxia. *Blood, 9*:721, 1954.
5. Borsook, H.A., Graybiel, A., Keighley, G., and Windsor, E.: Polycythemic response in normal adult rats to a non-protein plasma extract from anemic rabbits. *Blood, 9*:734, 1954.
6. Erslev, A.J.: Humoral regulation of red cell production. *Blood, 8*:349, 1953.
7. Gordon, A.S., Piliero, S.J., Kleinberg, W., and Freedman, H. H.: A plasma extract with erythropoietic activity. *Proc Soc Exp Biol Med, 86*:255, 1954.
8. Jacobson, L.O., Plzak, L., Fried, W., and Goldwasser, E.: Plasma factor(s) influencing red cell production. *Nature, 177*:1240, 1956.
9. Jacobson, L.O.: Sites of formation of erythropoietin. In Jacobson, L.O., and Doyle, M. (Eds.): *Erythropoiesis*. New York, Grune & Stratton, 1962, p. 69.

10. Gallagher, N.I., McCarthy, J.M., and Lange, R.D.: Erythropoietin production in uremic rabbits. *J Lab Clin Med, 57*:281, 1961.
11. Gordon, A.S., Cooper, G.W., and Zanjani, E.D.: The kidney and erythropoiesis. *Semin Hematol, 4*:337, 1967.
12. Osnes, S.: Experimental study of an erythropoietic principle produced in the kidney. *Br Med J, 2*:650, 1959.
13. Demopoulos, H.B., Highman, B., Ahland, P.D., Gerving, M.A., and Kaley, G.: Effects of high altitude on granular juxtaglomerular cells and their possible role in erythropoietin production. *Am J Pathol, 46*:497, 1965.
14. Hirashima, K., and Takaku, F.: Experimental studies on erythropoietin. II. The relationship between juxtaglomerular cells and erythropoietin. *Blood, 20*:1, 1962.
15. Bourgoignie, J.J., Gallagher, N.I., Perry, H.M., Jr., Kurz, L., Warnecke, M.A., and Donati, R.M.: Renin and erythropoietin in normotensive and in hypertensive patients. *J Lab Clin Med, 71*:523, 1968.
16. Donati, R.M., Bourgoignie, J.J., Kuhn, C., Gallagher, N.I., and Perry, H.M., Jr.: Dissociation of circulating renin and erythropoietin in rats. *Circ Res, 22*:91, 1968.
17. Jacobson, L.O., Goldwasser, E., Plzak, L., and Fried, W.: Studies on erythropoiesis. IV. Reticulocyte response of hypophysectomized and polycythemic rodents to erythropoietin. *Proc Soc Exp Biol Med, 94*:243, 1957.
18. Filmanowicz, E., and Gurney, C.W.: Studies on erythropoiesis. XVI. Response to a single dose of erythropoietin in polycythemic mouse. *J Lab Clin Med, 57*:65, 1961.
19. Alpen, E.L., Cranmore, D., and Johnston, M.E.: Early observation on the effects of blood loss. In Jacobson, L.O., and Doyle, M. (Eds.): *Erythropoiesis*. New York, Grune & Stratton, 1962, p. 184.
20. Hodgson, G.: Erythrocyte Fe^{59} uptake as a function of bone marrow dose injected in lethally irradiated mice. *Blood, 19*:460, 1962.
21. Nakao, K., Takaku, F., Fujioka, S., and Sassa, S.: The effect of erythropoietin on hematopoietic organs of the polycythemic mouse. *Blood, 27*:537, 1966.
22. Reissmann, K.R.: Selective eradication of erythropoiesis by actinomycin D as the result of interference with hormonally controlled effector pathway of cell differentiation. *Blood, 28*:201, 1966.
23. Linman, J.W.: Factors controlling hemopoiesis: Erythropoietic effects of "anemic" plasma. *J Lab Clin Med, 59*:249, 1962.
24. Whitcomb, W.H., and Moore, M.: The physiological significance of an erythropoietic inhibitor factor appearing in plasma subsequent to hypertransfusion. *Ann NY Acad Sci, 149*:462, 1968.
25. Whitcomb, W.H., Moore, M., and Rhoada, J.P.: Influence of polycythemic and anemic plasma on erythrocyte iron incorporation in the plethoric hypoxic mouse. *J Lab Clin Med, 73*:584, 1969.

26. Gallagher, N.I., Hagan, D.Q., McCarthy, J.M., and Lange, R.D.: Response of starved rats and polycythemic rats to graded doses of erythropoietin. *Proc Soc Exp Biol Med, 106*:127, 1961.
27. Fried, W., Plzak, L., Jacobson, L.O., and Goldwasser, E.: Erythropoiesis. II. Assay of erythropoietin in hypophysectomized rats. *Proc Soc Exp Biol Med, 92*:203, 1956.
28. Jacobson, J.O., Goldwasser, E., Plzak, L., and Fried, W.: Studies on erythropoiesis. IV. Reticulocyte response of hypophysectomized and polycythemic rodents to erythropoietin. *Proc Soc Exp Biol Med, 94*:243, 1957.
29. Cotes, P.M., and Bangham, D.R.: Bio-assay of erythropoietin in mice made polycythemic by exposure to air at a reduced pressure. *Nature, 191*:1065, 1961.
30. Lange, R.D., Simmons, M.L., and McDonald, T.P.: Use of silicone rubber membrane enclosures for preparation of erythropoietin assay mice. *Ann NY Acad Sci, 149*:34, 1968.
31. Krantz, S.B., and Jacobson, L.O.: *Erythropoietin and the Regulation of Erythropoiesis*. Chicago, University of Chicago Press, 1970, p. 176.
32. Modan, B.: Polycythemia: A review of epidemiological and clinical aspects. *J Chronic Dis, 18*:605, 1965.
33. Levin, W.C., and Alperin, J.B.: An endocrinologic classification of polycythemia based on the production of erythropoietin. *Am J Med Sci, 256*:131, 1968.
34. Gallagher, N.I., and Donati, R.M.: Inappropriate erythropoietin elaboration. *Ann NY Acad Sci, 149*:528, 1968.
35. Lange, R.D., and Pavlovic-Kentera, V.: Erythropoietin. In Moore, C.V., and Brown, E.B. (Eds.): *Progress in Hematology*. New York, Grune & Stratton, 1964, vol. 4, p. 72.
36. Alexanian, R.: Urinary excretion of erythropoietin in normal men and women. *Blood, 28*:344, 1966.
37. Alexanian, R.: Erythropoietin excretion in man following androgens. *Blood, 28*:1007, 1966.
38. Alexanian, R.: Erythropoietin and erythropoiesis in anemic man following androgens. *Blood, 33*:564, 1969.
39. Adamson, J.W.: The erythropoietin/hematocrit relationship in normal and polycythemic man: Implications of marrow regulation. *Blood, 132*:597, 1968.
40. Adamson, J.W., and Finch, C.A.: Erythropoietin and the regulation of erythropoiesis in DiGulielmo's syndrome. *Blood, 36*:390, 1970.

AUTHOR INDEX

SUBJECT INDEX